XI Mexican School on Gravitation and Mathematical Physics 2016

Quantum Gravity: Schemes, Models and Phenomenology

Journal of Physics: Conference Series Volume 1030

Playa del Carmen, Mexico
5 - 9 December 2016

Editors:

Hector Hernandez
Guillermo Chacon
Miguel Sabido

ISBN: 978-1-5108-6490-0
ISSN: 1742-6588

Printed from e-media with permission by:

Curran Associates, Inc.
57 Morehouse Lane
Red Hook, NY 12571

Some format issues inherent in the e-media version may also appear in this print version.

Printed by Curran Associates, Inc. (2018)

For permission requests, please contact the Institute of Physics
at the address below.

Institute of Physics
Dirac House, Temple Back
Bristol BS1 6BE UK

Phone: 44 1 17 929 7481
Fax: 44 1 17 920 0979

techtracking@iop.org

Additional copies of this publication are available from:

Curran Associates, Inc.
57 Morehouse Lane
Red Hook, NY 12571 USA
Phone: 845-758-0400
Fax: 845-758-2633
Email: curran@proceedings.com
Web: www.proceedings.com

Table of contents

Volume 1030

XI Mexican School on Gravitation and Mathematical Physics, "Quantum Gravity: schemes, models and phenomenology"

5–9 December 2016, Playa del Carmen, Quintana Roo, Mexico

Accepted papers received: 9 May 2018
Published online: 12 June 2018

Preface

Papers

IOP Conf. Series: Journal of Physics: Conf. Series **1030** (2018) 011001 doi:10.1088/1742-6596/1030/1/011001

Preface

The collection of papers in this volume was presented during the XI Mexican School on Gravitation and Mathematical Physics, which was held in Playa del Carmen, Quintana Roo, México, December 5-9, 2016. The Mexican School on Gravitation and Mathematical Physics is a series of conferences sponsored by the Mexican Physical Society that started in 1994 with the purposes of discussing and exchanging current ideas in gravitational physics. Each Mexican School has been devoted to a particular subject, and these have included supergravity, branes, black holes, the early Universe, observational cosmology, quantum gravity and numerical relativity.

In this occasion the theme of the school was Quantum Gravity: schemes, models and phenomenology, which focused on the discussion of quantum aspects of gravity. Following our previous Schools, world leaders in the field were invited to give courses and plenary lectures. More specialized talks were also presented in parallel sessions, and some of them have been included in these proceedings.
The contributions in this volume have been reviewed and represent some of the courses, plenary talks and contributed talks presented during our XI School. We are indebted to the contributors of these proceedings as well as to the rest of the participants in our Mexican School all for making of it a complete success.

As for financial support we should mention the Mexican National Science and Technology Council (CONACyT), the Mexican Physical Society (SMF), as well as several Institutions including: Centro de Investigación y Estudios Avanzados (CINVESTAV), Universidad Autónoma Metropolitana Iztapalapa (UAM-I), Universidad de Guanajuato (UG), and Universidad Nacional Autónoma de México (UNAM).

Editors:
Hector Hernandez, Guillermo Chacón and Miguel Sabido

IOP Conf. Series: Journal of Physics: Conf. Series **1030** (2018) 011002 doi:10.1088/1742-6596/1030/1/011002

Peer review statement

All papers published in this volume of *Journal of Physics: Conference Series* have been peer reviewed through processes administered by the proceedings Editors. Reviews were conducted by expert referees to the professional and scientific standards expected of a proceedings journal published by IOP Publishing.

IOP Conf. Series: Journal of Physics: Conf. Series **1030** (2018) 012001 doi:10.1088/1742-6596/1030/1/012001

An algorithm for quantum gravity phenomenology

Yuri Bonder

E-mail: bonder@nucleares.unam.mx
Instituto de Ciencias Nucleares, Universidad Nacional Autónoma de México
Apartado Postal 70-543, Ciudad de México, 04510, México

Abstract. Quantum gravity phenomenology is the strategy towards quantum gravity where the priority is to make contact with experiments. Here I describe what I consider to be the best procedure to do quantum gravity phenomenology. The key step is to have a generic parametrization which allows one to perform self-consistency checks and to deal with many different experiments. As an example I describe the role that the Standard Model Extension has played when looking for Lorentz violation.

1. Introduction

Finding a fully consistent theory of quantum gravity is one of the biggest challenges in physics (for an introductory review see Ref. [1]). One obstacle to build this theory is the lack of experimental data. The program generically known as quantum gravity phenomenology is comprised by several research strategies sharing the willingness to make contact with experiments [2]. Basically, the starting point for any of this strategies is to make an educated guess on how can a quantum gravity effect arise at testable scales. Then, some mathematical expressions are given containing free parameters that must be constrained by comparing with experiments and, for which, a nonzero value can be considered to be evidence of new physics.

The claim is that, by refining the methods that are usually utilized, we could pursue the goals of quantum gravity phenomenology more efficiently. I introduce the proposal as an algorithm whose flowchart is presented in figure 1. I first describe this algorithm; then, as an example, I discuss how it has been implemented in the case of Lorentz violation.

2. The algorithm

The first step of the proposed algorithm is to narrow down the search for new physics. Of course, given that we know essentially nothing about the quantum gravity regime, there is no criteria that could help us make this decision. However, one can classify the possible research avenues. The proposal is to do so in terms of the principles of the current paradigm. That is, to select one of these principles and then look for deviations from conventional physics where this particular principle does not hold. For the case at hand, a list of such principles includes local Lorentz invariance, the torsion-free hypothesis, the fact that the spacetime dynamics is determined by the Einstein-Hilbert action, and all the principles of Quantum Mechanics.

The use of principles as a starting point is a natural choice because, as happened in every scientific revolution, a dramatic change in this part of the theory must occur [3]. Note, however, that this classification is not problem free. For example, there is no consensus on what the principles underlying current physics are or how should they be enunciated. In addition, there

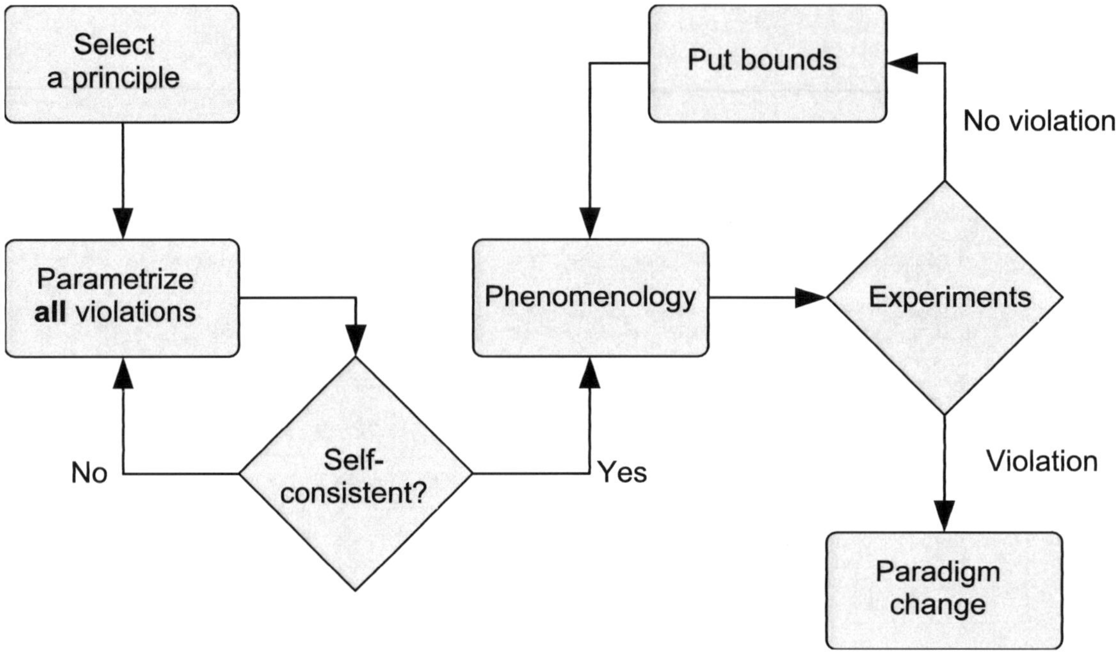

Figure 1. The proposed algorithm for quantum gravity phenomenology.

could be some principles that are not independent. However, using the most basic structure in the theory minimizes the risks of propagating ambiguities into the following steps.

Once a principle is chosen it is extremely useful to parametrize all possible deviations from conventional physics where this principle is violated. Looking for *all* modifications reflects the fact that we have no clues about the nature of quantum gravity. I would like to stress that the quantum gravity phenomenology literature is filled with examples where the opposite view point is taken.

Moreover, a generic parametrization can encode all relevant experimental results as conditions on its parameters. In this sense the parametrization provides us with a tool to compare different experiments. Even more, it can motivate new experiments since the mathematical expressions can guide our physical intuition, as often occurs in physics. Finally, the fact that the parametrization is done on top of conventional physics is useful to keep track of the physical meaning of the mathematical objects at hand.

It is important to stress that, in practice, it is hard to write this general parametrization. What typically happens is that one uses additional criteria to select a subspace of the parametrization and tries to be general under such a restriction. Something useful to define these subspaces is the fact that the structure of conventional physics permeates into the parametrization, providing a classification of all possible new effects in terms of 'sectors.' For example, one can decide to focus only on an extension to Quantum Electrodynamics, or General Relativity, and so on. Moreover, the parametrization may have an intrinsic hierarchy based, essentially, on dimensional analysis, and which may be used to argue what modifications could dominate. Once the generic parametrization is built, it is possible to verify if it is internally consistent. This is a paramount step, which is often ignored, that can discard parts of the parametrization.

Finally, once the self-consistent parametrization is obtained, it is time to compare with experiments. For that purpose, and given a concrete experiment, it is useful to select the relevant sector and to perform the corresponding approximations. Clearly, the resulting expressions carry the parameters and, if no evidence of violations is shown, bounds on such parameters can be placed. In principle, one would then try to improve the experiment and continue testing, improving the limits on the parameters on each cycle. Of course, there is always the chance to detect a signal that leads to a paradigm change, which is the ultimate goal of the program.

It should be mentioned that, even if no signs of new physics are uncovered, the results of these analyses are extremely valuable. This is because it motivates new experiments, which can have unexpected outcomes, but, more importantly, since the bounds on the parameters can be used to rule out different theories and to have a quantitative measure of the validity of the principle that is tested.

3. An example: Lorentz violation

The community that best implements the method described above revolves around the so-called Standard Model Extension (SME), which is a parametrization for violations of local Lorentz invariance [4, 5]. Recall that local Lorentz invariance states that locally and in free fall there are no preferred directions associated with spacetime. Now, the SME is built in the effective field theory context. Thus, it has an action containing the conventional-physics action plus new terms that are not Lorentz invariant [6]. These new terms have a parameter for Lorentz violation, called an SME coefficient, which is coupled to conventional fields. Also, the field content and the gauge symmetries of conventional physics are unchanged. Remarkably, the SME has triggered many theoretical and experimental works, some of which are described next.

3.1. Self-consistency checks

One of the earliest theoretical analysis done within the SME has to do with field redefinitions [7]. It was noted that such redefinitions can be used to cancel some terms. These studies explained why some coefficients did not show up in the phenomenological expressions and are now a key part of any phenomenological analysis.

Another example is related to the fact that, when Lorentz violation is studied on a dynamical spacetime, the metric equation of motion contains the Einstein tensor, which is divergence free by virtue of the Bianchi identity. Thus, the divergence of this equation severely restricts the values of the SME coefficients. The most popular solution within the SME community is to consider that Lorentz violation arises spontaneously [5]. That is, that there are action terms for the coefficients with potentials that favor those configurations where preferred directions are selected.

There are methods to avoid using a particular action term for each SME coefficient [8]. However, these methods can only be implemented perturbatively and cannot be used to look for the effects of all the SME coefficients [9]. Note in passing that gravitational effects, for tabletop experiments, can be introduced without having to use spontaneous Lorentz violation by using a uniformly accelerated frame [10].

The last example I want to mention has to do with the Cauchy problem. A desirable property of physical theories is that, given proper initial data, they make unique predictions. When the theories also satisfy particular causality and continuity relations between initial-data changes and solution changes, they are said to have a well-posed Cauchy problem [11]. Now, the most popular models of spontaneous Lorentz violation, the so-called Bumblebee Models, consist of a vector field with a Maxwell-like kinetic term and potentials for the field square that generate spontaneous Lorentz violation [12]. It has been shown that these models, in flat spacetime and coupled to a scalar field that plays the role of matter, do have a Hamiltonian that respects the constraints [13]. However, it was also shown that the initial data, as required by the Dirac analysis [14], does not determine the evolution uniquely [13]. Concretely, given a set of initial data, it is possible to find several physically-inequivalent evolutions that are all consistent with the same initial data. This is an example of a consistency analysis that can be used to discard some SME parts.

An easy exit to the objection described in the last paragraph is to simply change the kinetic term. The problem, however, is that most non-Maxwell kinetic terms depend on the spacetime connection. Therefore, by fixing the Cauchy problem for the vector field one may inadvertently damage that of the metric. This should be studied carefully but it is appealing as it can restrict further these models for spontaneous Lorentz violation.

3.2. Experimental results

The most exciting feature of the SME is the continuous flow of new experiments that look for Lorentz violation. These experiments are of various types; a partial list includes:

- Accelerator and collider experiments.
- Astrophysical observations.
- Vacuum birefringence and dispersion of light.
- Clock-comparison experiments.
- Laboratory gravity tests.
- Matter interferometry.
- Neutrino oscillations.
- Particle *vs.* antiparticle comparisons.
- Resonant cavities and lasers.
- Sidereal/annual variations of physical signals.

All the results from these experiments are collected in the so-called Data Tables for Lorentz Violation [15] that quote more that 150 experimental results. Remarkably, several of the empirical limits that have been obtained are transplanckian. In fact, in some cases the limits approach two powers of the ratio of the electroweak and the Planck scales, which is the number where one can naively expect quantum gravity effects to show up.

4. Conclusions

There is an active community working on quantum gravity phenomenology for which the main goal is to search for experimental clues of quantum gravity. Here a procedure to work in the area is presented. The basic nonstandard steps are to have a generic parametrization of all possible violations of a principle and to do self-consistency checks. This strategy has been implemented mainly by the Standard Model Extension program that has motivated many interesting experimental and theoretical studies.

It should be said that, thus far, there is no compelling empirical evidence of quantum gravity. This only means that there is a lot of work to be done and, ideally, many studies will follow the steps outlined here. I am confident that, if we keep testing for the principles of current physics, we will eventually uncover new physics.

Acknowledgements

I thank Alan Kostelecký for many discussion on this topic and I acknowledge financial support from UNAM-DGAPA-PAPIIT Grant No. IA101116 and Red Temática CONACyT "Física de Altas Energías."

IOP Conf. Series: Journal of Physics: Conf. Series **1030** (2018) 012001 doi:10.1088/1742-6596/1030/1/012001

References

[1] A. Ashtekar, Current Science **89**, 2064 (2005).
[2] G. Amelino-Camelia, Liv. Rev. Rel. **16**, 5 (2013).
[3] T. S. Kuhn, *The Structure of Scientific Revolutions*, (University of Chicago Press, Chicago, 1996).
[4] D. Colladay and V.A. Kostelecký, Phys. Rev. D **55**, 6760 (1997); Phys. Rev. D **58**, 116002 (1998).
[5] V.A. Kostelecký, Phys. Rev. D. **69**, 105009 (2004).
[6] V.A. Kostelecký and R. Potting, Phys. Rev. D **51**, 3923 (1995).
[7] V.A. Kostelecký and M. Mewes, Phys. Rev. D **66**, 056005 (2002); D. Colladay and P. McDonald, J. Math. Phys. **43**, 3554 (2002); Q.G. Bailey and V.A. Kostelecký, Phys. Rev. D **70**, 076006 (2004); R. Lehnert, Phys. Rev. D **74**, 125001 (2006); B. Altschul, J. Phys. A **39**, 13757 (2006).
[8] Q.G. Bailey and V.A. Kostelecký, Phys. Rev. D **74**, 045001 (2006).
[9] Q.G. Bailey, V.A. Kostelecký, and R. Xu, Phys. Rev. D **91**, 022006 (2015); Y. Bonder, Phys. Rev. D **91**, 125002 (2015).
[10] Y. Bonder, Phys. Rev. D **88**, 105011 (2013).
[11] R.M. Wald, *General Relativity*, (University of Chicago Press, Chicago, 1984), chapter 10.
[12] R. Bluhm and V.A. Kostelecký, Phys. Rev. D **71**, 065008 (2005).
[13] Y. Bonder and C.A. Escobar, Phys. Rev. D **93**, 025020 (2016).
[14] P.A.M. Dirac, *Lectures on Quantum Mechanics*, (Dover Publications, New York, 2001).
[15] V.A. Kostelecký and N. Russell, Rev. Mod. Phys. **83**, 11 (2011).

IOP Conf. Series: Journal of Physics: Conf. Series **1030** (2018) 012002 doi:10.1088/1742-6596/1030/1/012002

Moyal product for $(n-1)$-forms within the covariant Hamiltonian formalism for fields

Jasel Berra–Montiel[1], Alberto Molgado[1,2] and David Serrano-Blanco[1]

[1] Facultad de Ciencias, Universidad Autonoma de San Luis Potosi
Av. Salvador Nava S/N Zona Universitaria, San Luis Potosi, SLP, 78290 Mexico
[2] Dual CP Institute of High Energy Physics, Mexico

E-mail: jberra@fc.uaslp.mx, molgado@fc.uaslp.mx, davidserrano@fc.uaslp.mx

Abstract. In these brief notes we analyze an associative product for $(n-1)$-forms in the polymomentum phase space which is necessary in order to complete the deformation quantization scheme within the context of the multisymplectic formalism. In particular, we identify certain conditions that the proposed product must follow to generate a Moyal bracket induced from the Gerstenhaber bracket naturally emerging within this context.

1. Introduction

The process in which a quantum theory is constructed starts in most cases as a classical system described in configuration space and then taken to its phase-space formulation. This map requires a privileged direction in the physical space to be chosen and singled out from the other coordinates in a process known as foliation of spacetime. In a physical theory this direction is often chosen as the time coordinate. A problem arises though, if one considers the principle of general covariance which states that there cannot be a privileged direction of the physical space in which a relativistic theory is described. Conventional quantization schemes based on a Hamiltonian description of a theory depend on the election of time as a special coordinate. There is, however, a less conventional formalism to get the Hamiltonian picture from the Lagrangian scheme in a completely covariant way proposed more than half century ago [1]. Indeed, while the Hamiltonian formalism possesses an intrinsic geometric symplectic structure, the covariant Hamiltonian scheme requires a generalization of this framework, commonly referred to as the multisymplectic structure, which emerges by modifying the Legendre transformation when requiring it to treat all directions of spacetime in the same way. The main purpose of this paper is to analyze the possibility of applying the scheme of deformation quantization [2] to the polysymplectic formalism. In particular we aim to study the promotion of the Gerstenhaber bracket, naturally emerging in the polysymplectic formalism, to a Moyal bracket. Our work thus focuses precisely in characterizing an associative product for $(n-1)$-forms, commonly associated to observables within the polysymplectic formalism.

The rest of this work is presented as follows: In section 2 we briefly review the polysymplectic formalism. In Section 3 we analyze the polysymplectic structure for $(n-1)$-Hamiltonian forms and propose a Moyal bracket for them. Finally, Section 4 consists of some concluding remarks.

2. Polysymplectic Geometry

In this section we succinctly review the multisymplectic approach to the De Donder-Weyl formalism for classical field theory [3, 4]. Let us start by considering a fiber bundle (E, π, M), with M an n-dimensional smooth manifold with local coordinates $\{x^\mu\}_{\mu=1}^n$, E a smooth manifold of dimension m with local coordinate functions $\{y^a\}_{a=1}^m$ and $\pi : E \to M$ the standard projection map. Let $\phi : M \to E$ be a local section around a point $p \in M$, so that its local components are given by $\phi^a := y^a \circ \phi$. We will denote the space of local sections of π around the point p as $\Gamma_p(\pi)$. Now, let us define the configuration space as the first jet manifold J^1E of π, whose elements $j^1_p\phi \in J^1E$, for each point p, and every local section ϕ, have the local representation $(x^\mu, \phi^a, \phi^a_\mu)$, where $\phi^a_\mu := \partial\phi^a/\partial x^\mu$ are commonly known as the derivative coordinates [3, 5]. The behaviour of a physical system is encoded by the action functional $S : \Gamma_p(\pi) \to \mathbb{R}$, given in terms of the Lagrangian density $\mathcal{L} : J^1E \to \mathbb{R}$ such that $\mathcal{L}(j^1_p\phi) = \mathcal{L}(x^\mu, \phi^a, \phi^a_\mu)$, by $\mathcal{S}[\phi] := \int_{\mathcal{C}} \mathcal{L}(j^1_p\phi)\, V$, for any $\phi \in \Gamma_p(\pi)$, where V denotes the local volume element of M and $\mathcal{C}$ is an appropriate submanifold of M on which the integration is performed. Let us define the quantities $\pi^\mu_a := \partial\mathcal{L}/\partial\phi^a_\mu$, called the polymomenta. The triple $(x^\mu, \phi^a, \pi^\mu_a)$ thus defines a local coordinate system for a new manifold $\mathcal{P}$, known in this context as the *polymomentum phase-space*. The Lagrangian density $\mathcal{L}$ is associated to a smooth function H_{DW} in $\mathcal{P}$ by the map $\mathbb{L} : C^\infty(J^1E) \longrightarrow C^\infty(\mathcal{P})$, called the covariant Legendre transformation, such that $H_{DW}(x^\mu, \phi^a, \pi^\mu_a) := \mathbb{L}(\mathcal{L}) = \pi^\mu_a\, \partial_\mu\phi^a - \mathcal{L}(x^\mu, \phi^a, \pi^\mu_a)$, where the symbol ∂_μ stands for the derivative with respect to the base space coordinates x^μ. The function $H_{DW} : \mathcal{P} \longrightarrow \mathbb{R}$ is called the *De Donder-Weyl Hamiltonian* [3].

As it is well known, the polymomentum phase-space $\mathcal{P}$ is also endowed with a canonical n-form induced from the Lagrangian Cartan form Θ_L with the local representation $\Theta_{DW} = \pi^\mu_a\, d\phi^a \wedge V_\mu - H_{DW} V$, known as the Poincaré-Cartan form [3, 6]. This local representation explicitly shows the decomposition of Θ_{DW} into two terms, its vertical and horizontal parts, respectively. As it has been shown before, the behaviour of a physical system may be encoded only in the vertical term of the Poincaré-Cartan form [3]. Now, given an arbitrary p-form in the polymomentum phase-space, namely, $\overset{p}{F} = \frac{1}{p!} F_{M_1\cdots M_p} dz^{M_1} \wedge \cdots \wedge dz^{M_p}$, where $z^M := (x^\mu, \phi^a, \pi^\mu_a)$ denotes the basis of $\mathcal{P}$, then we define its vertical derivative by $d^V \overset{p}{F} := (1/p!)\partial_v F_{M_1\cdots M_p} dz^v \wedge dz^{M_1} \wedge \cdots \wedge dz^{M_p}$, where $z^v := (\phi^a, \pi^\mu_a)$ stands for the vertical coordinates in $\mathcal{P}$ [4]. In this way, by taking the vertical derivative of Θ^V_{DW} we may define $\Omega_{DW} := d^V \Theta^V_{DW} = d\pi^\mu_a \wedge d\phi^a \wedge V_\mu$, commonly referred to as the polysymplectic form [4]. This $(n+1)$-form defines an analogue of the usual symplectic structure in the polymomentum phase-space. We will call $\overset{p}{X}$ a Hamiltonian multivector field of degree p, if there exists a unique horizontal form $\overset{n-p}{F}$ satisfying $\overset{p}{X} \lrcorner\, \Omega_{DW} = d^V \overset{n-p}{F}$. Thus $\overset{n-p}{F}$ is called the Hamiltonian form associated to the Hamiltonian multivector field $\overset{p}{X}$. Let $\overset{p}{F}$, and $\overset{q}{G}$ be p and q-Hamiltonian forms, respectively, and let $\overset{n-p}{X}_F$ and $\overset{n-q}{X}_G$, their associated Hamiltonian fields. The map given by

$$\{\![\, \overset{p}{F}, \overset{q}{G} \,]\!\} := (-1)^{n-p} \overset{n-p}{X}_F \lrcorner \overset{n-q}{X}_G \lrcorner \Omega_{DW} \,, \tag{1}$$

is called the Gerstenhaber Bracket (GB) [4, 7]. This bracket structure results graded-commutative, and satisfies Jacobi identity and Leibniz rule under the co-exterior product, $\bullet$, given by $\overset{p}{F} \bullet \overset{q}{G} := *^{-1}(*\overset{p}{F} \wedge *\overset{q}{G})$, with $*$ being the Hodge dual defined over M, and thus it is a graded-Poisson bracket. In particular, one may show that the GB closes for $(n-1)$-forms. Besides, this co-exterior product induces a derivative operator called the total co-exterior differential over horizontal forms [4]. Hence, the co-exterior differential of a Hamiltonian form $\overset{p}{F}$ can be written as $\mathrm{d} \bullet \overset{p}{F} = \{\![\, H_{DW}, \overset{p}{F} \,]\!\} + d^h \bullet \overset{p}{F}$. Finally, using the canonical commutation relations, $\{\![\, \pi^\mu_a V_\mu, \phi^b \,]\!\} = \delta^b_a$, $\{\![\, \pi^\mu_a V_\mu, \phi^b V_\nu \,]\!\} = \delta^b_a V_\nu$ and $\{\![\, \pi^\mu_a, \phi^b V_\nu \,]\!\} = \delta^b_a \delta^\mu_\nu$, we obtain the De Donder-Weyl equations of motion [3]

$$\partial_\mu \phi^a = \{\![\, H_{DW}, \phi^a V_\mu \,]\!\} = \frac{\partial H_{DW}}{\partial \pi^\mu_a} \,, \qquad \partial_\mu \pi^\mu_a = \{\![\, H_{DW}, \pi^\mu_a V_\mu \,]\!\} = -\frac{\partial H_{DW}}{\partial \phi^a} \,. \tag{2}$$

It is worth noting that whenever we specifically consider $M = \mathbb{R}$ and $E = \mathbb{R}^n$, then all the above formalism reduces correspondingly to both standard Lagrangian and Hamiltonian formalisms for particle mechanics.

3. Moyal star-product in the polymomentum phase space

Our main aim is to apply the method of deformation quantization to the multisymplectic framework. As it is well known, the Moyal star-product in the standard symplectic case can be written in a differential representation [2] as $f \star g = f \exp\left[(i\hbar/2)\left(\overleftarrow{\partial}_{q^i}\overrightarrow{\partial}_{p_i} - \overleftarrow{\partial}_{p_i}\overrightarrow{\partial}_{q^i}\right)\right] g$, where the arrows point to the function they act on. Also, within this case, we may define the Moyal bracket as $\{\!\{\, f, g \,\}\!\} := (f \star g - g \star f)$. This structure related to the Poisson bracket within the phase-space (together with Weyl and Wigner transforms) constitute the basic elements of the deformation quantization scheme [8]. Thus, the first step towards a deformation quantization scheme based on the DW-Hamiltonian, is to construct an analogue star-product to this Moyal star-product. In particular, as $(n-1)$-forms play a privileged role under the GB structure (1) we will focus on them. Therefore, in local coordinates the GB may be found by means of the product $\overset{n-p}{X}_F \lrcorner \overset{n-q}{X}_G \lrcorner \Omega$, which explicitly reads

$$\{\![\, \overset{n-1}{F}, \overset{n-1}{G} \,]\!\} = \overset{n-1}{F} (\overleftarrow{\Delta}^{\mu} \overleftarrow{\partial}_a \overrightarrow{\partial}^a_\mu - \overleftarrow{\partial}^a_\mu \overrightarrow{\partial}_a \overrightarrow{\Delta}^\mu) \overset{n-1}{G} \,, \tag{3}$$

where the operator $\Delta^\mu := (1/(n-1)!)\epsilon^{\mu\mu_1\cdots\mu_{n-1}}\partial_{\mu_1\cdots\mu_{n-1}}$, stands for the horizontal dual element to V_μ (that is, $\Delta^\mu \lrcorner V_\nu = \delta^\mu_\nu$). This operator acts by appropriate contraction, either $\overrightarrow{\Delta}^\mu \overset{n-1}{F} = \Delta^\mu \lrcorner \overset{n-1}{F}$ or $\overset{n-1}{F} \overleftarrow{\Delta}^\mu = \Delta^\mu \lrcorner \overset{n-1}{F}$. In this way, by defining the differential operator $\overleftrightarrow{D} := \overleftarrow{\Delta}^{\mu} \overleftarrow{\partial}_a \overrightarrow{\partial}^a_\mu - \overleftarrow{\partial}^a_\mu \overrightarrow{\partial}_a \overrightarrow{\Delta}^\mu$, the GB 3 can be expressed as $\overset{n-1}{F} \overleftrightarrow{D} \overset{n-1}{G}$. Next, we note that if we want to construct a second order differential operator from (3) by bare composition of the operator $\overleftrightarrow{D}$, we would expect terms involving $\overrightarrow{\Delta}^\mu \overrightarrow{\Delta}^\nu$, and so on, however, once acting over an $(n-1)$-form, this would yield terms as $\vec{\Delta}^{\mu}\tilde{F}^{\nu}$, which are not defined over the space of horizontal forms any more. To deal with this inconvenience, we need to define a product between this operators so that the degree is maintained. We start by assuming the existence of this product, and let us denote it by $\circ$. Thus, we define the operator $\overleftrightarrow{D}^2 := \overleftrightarrow{D} \circ \overleftrightarrow{D}$, so that $\overset{n-1}{F}\overleftrightarrow{D}^2\overset{n-1}{G} = \overset{n-1}{F}(\overleftarrow{\Delta}^{\mu} \circ \overleftarrow{\Delta}^{\nu} \overleftarrow{\partial}_a \overleftarrow{\partial}_b \vec{\partial}^a_{\mu} \vec{\partial}^b_{\nu} - \overleftarrow{\partial}_a \overleftarrow{\partial}^b_{\nu} \overleftarrow{\Delta}^{\mu} \circ \vec{\Delta}^{\nu} \vec{\partial}^a_{\mu} \vec{\partial}_b - \overleftarrow{\partial}_b \overleftarrow{\partial}^a_{\mu} \overleftarrow{\Delta}^{\nu} \circ \vec{\Delta}^{\mu} \vec{\partial}^b_{\nu} \vec{\partial}_a + \overleftarrow{\partial}^a_{\mu} \overleftarrow{\partial}^b_{\nu} \vec{\partial}_a \vec{\partial}_b \vec{\Delta}^{\mu} \circ \vec{\Delta}^{\nu})\overset{n-1}{G}$. It is clear from this expression that we need to define separate cases for the product $\circ$, namely, $\overleftarrow{\Delta}^{\mu} \circ \overleftarrow{\Delta}^{\nu} := \overleftarrow{\Delta}^{\sigma} \lambda^{\mu\nu}_{\sigma}$, $\vec{\Delta}^{\mu} \circ \vec{\Delta}^{\nu} := \vec{\Delta}^{\sigma} \lambda^{\mu\nu}_{\sigma}$ and $\overleftarrow{\Delta}^{\mu} \circ \vec{\Delta}^{\nu} := \frac{1}{2}\lambda^{\mu\nu}_{\sigma}(\overleftarrow{\Delta}^{\sigma} + \vec{\Delta}^{\sigma})$, where $\lambda^{\mu\nu}_{\sigma}$ are constants to be fixed, and such that the result in every term of $\overset{n-1}{F}\overleftrightarrow{D}^2\overset{n-1}{G}$ is always an $(n-1)$-form. With this definitions, the previous result is then simply written as

$$\overset{n-1}{F}\overleftrightarrow{D}^2\overset{n-1}{G} = \lambda^{\mu\nu}_{\sigma}(\partial_a \partial_b \tilde{F}^{\sigma} \partial^a_{\mu} \partial^b_{\nu} \tilde{G}^{\rho} - \partial_a \partial^b_{\nu} \tilde{F}^{\sigma} \partial^a_{\mu} \partial_b \tilde{G}^{\rho} - \partial_a \partial^b_{\nu} \tilde{F}^{\rho} \partial^a_{\mu} \partial_b \tilde{G}^{\sigma} + \partial^a_{\mu} \partial^b_{\nu} \tilde{F}^{\rho} \partial_a \partial_b \tilde{G}^{\sigma})V_{\rho}. \quad (4)$$

Here, for simplicity, we used the definitions $\tilde{F}^{\sigma} := (1/(n-1)!)\epsilon^{\nu_1 \cdots \nu_q \sigma} F_{\nu_1 \cdots \nu_q}$, where ϵ and $F_{\nu_1 \cdots \nu_q}$ stand for the alternating symbol and the components of the $(n-1)$-form F, respectively, and also $V_{\mu} := \partial_{\mu} \lrcorner V$. Noticing that the derivatives with respect to the polymomenta commute, we need to impose the condition that the constants $\lambda^{\mu\nu}_{\sigma}$ are symmetric in the upper indices, making the product $\circ$, a commutative operation. This last expression yields an $(n-1)$-form, and furthermore, it is clear that repeated applications of $\overleftrightarrow{D}$ also result in an $(n-1)$-form. We take advantage of the arbitrariness of the constants $\lambda^{\sigma}_{\mu\nu}$ to make this map obey Leibniz product rule under the GB. We see that a sufficient requirement for the $\circ$ product to obey Leibniz product rule is the identity $T^{\mu}_{\nu}\lambda^{\nu}_{\alpha\beta}\tilde{G}^{\alpha} = T^{\alpha}_{\nu}\lambda^{\mu}_{\alpha\beta}\tilde{G}^{\nu}$, for an arbitrary quantity T^{μ}_{ν}. This imposition on the constants also ensure that the $\circ$ product is associative. Having constructed an associative product between $(n-1)$-forms, we can now define the action of $\overleftrightarrow{D}^k$ when $k = 0$ as $\overset{n-1}{F}\overleftrightarrow{D}^0\overset{n-1}{G} := \overset{n-1}{F} \circ \overset{n-1}{G}$, so that the formal deformation of the product $\circ$, namely

$$\overset{n-1}{F} \star \overset{n-1}{G} := \overset{n-1}{F}\left[\sum_{k=0}^{\infty} \frac{1}{k!}\left(\frac{i\hbar}{2}\right)^k \overleftrightarrow{D}^k\right]\overset{n-1}{G} = \overset{n-1}{F} \circ \overset{n-1}{G} + \overset{n-1}{F}\left[\sum_{k=1}^{\infty} \frac{1}{k!}\left(\frac{i\hbar}{2}\right)^k \overleftrightarrow{D}^k\right]\overset{n-1}{G}, \quad (5)$$

defines a closed Moyal star-product over the space $(n-1)$-Hamiltonian forms. Even though this star-product is an actual deformation of the algebra of $(n-1)$-forms, there still exists a certain ambiguity in the definition of the $\circ$ product due to the above imposed identity on the λ's, as it may not be valid for any arbitrary set of constants.

4. Conclusions

The deformation quantization scheme may provide a feasible way to obtain the quantum analogues of the DW equations. In this note we analyzed the characteristics that an associative product for $(n-1)$ forms may have in order to introduce a deformation quantization for them. As discussed, we can define the classical observables as the Hamiltonian $(n-1)$-forms. This election can be justified by the fact that the Hamiltonian forms of this particular degree actually close under the GB. However, up until now, a closed and associative product between $(n-1)$-forms has not been found. The proposed product was constructed in order to fit the two requirements needed to give a GB-compatible algebra which can be deformed to obtain a Moyal star-product. This led to define the $\circ$ product of two $(n-1)$-forms as the $(n-1)$-form whose components are linear combinations of the product of the original component functions.

We found two main disadvantages in the definition of the $\circ$ product. The first is the possible lack of a geometrical meaning as this product between two $(n-1)$-forms gives no immediate geometric information. Nevertheless, the $\circ$ product could contain an intrinsic geometric structure given by the operators $\vec{\Delta}^{\mu}$ from which it has been defined as they could be associated to projection operators over the space of $(n-1)$-forms. The other disadvantage concerning the $\circ$ product is the condition that the constants $\lambda^{\sigma}_{\mu\nu}$ must satisfy. In this sense, the interpretation of $\vec{\Delta}^{\mu}$ as projection operators could bring additional support to our choice. As a projection operator is idempotent by definition, this is $\vec{\Delta}^2 = \vec{\Delta}$, the values of the λ constants must reduce to a generalization of a Kronecker delta, yielding an idempotent operator. The product we have proposed in these notes may be the first step towards a concrete algebraic structure needed for a deformation quantization of structures within the multisymplectic approach. Having obtained a Moyal product defined in polymomentum phase-space, the application of a deformation quantization scheme to concrete physical models can now be considered. Besides standard examples, such as scalar field and electrodynamics, the finite dimensional nature of the polymomentum phase-space makes reparametrization invariant field theories, such as gravity, perfect candidates to test our results on. A complete geometric characterization of this product is under progress, and will be addressed elsewhere.

Acknowledgments
AM acknowledges financial support from CONACYT-Mexico under project CB-2014-243433.

References

[1] T. De Donder, *Theorie Invariantive du Calcul des Variations* (Gauthier-Villars, Paris, 1935); H. Weyl, Ann. Math. (2) **36** 607–629 (1935); C. Carathéodory, Acta Sci. Math. (Szeged) **4** 193–216 (1929).

[2] L. A. Takhtajan, *Quantum Mechanics for Mathematicians*, Graduate Studies in Mathematics, Vol. 95 (American Mathematical Society, 2008); G. Esposito, G. Marmo, G. Sudarshan, *From Classical to Quantum Mechanics* (Cambridge University Press, 2004); C. K. Zachos, D. B. Fairlie and T. L. Curtright, *Quantum Mechanics on Phase Space* (World Scientific, 2005).

[3] G. Giachetta, L. Mangiarotti and G. Sardanashvily, *New Lagrangian and Hamiltonian methods in field theory* (World Scientific, 1997); M. J. Gotay, *A multisymplectic framework for classical field theory and the calculus of variations I. Covariant Hamiltonian formalism, in Mechanics*, Analysis and Geometry: 200 Years after Lagrange, M. Francaviglia, ed. (North Holland, Amsterdam, 1991) pp. 203–235; H. A. Kastrup, Phys. Rep. **101** 1–167 (1983).

[4] I. V. Kanatchikov, Aachen preprint PITHA 93/41 (1993), `arXiv:hep-th/9312162`; Rep. Math. Phys. **40** 225–234 (1997), `arXiv:hep-th/9710069`; Phys. Lett. **A283** 25–36 (2001), `arXiv:hep-th/0012084`; Rep. Math. Phys. **43** 157–70 (1999), `arXiv:hep-th/9810165`; In *Current Topics in Mathematical Cosmology* (Proc. Int. Seminar, Potsdam, Germany, 1998) M. Rainer and H.-J. Schmidt eds. (World Scientific) pp. 457-–67, `arXiv:gr-qc/9810076`.

[5] D. J. Saunders, *The Geometry of Jet Bundles*, London Mathematical Society lecture notes series Vol. 142 (Cambridge University Press, 1989).

[6] M. Forger and L. G. Gomes, Rev. Math. Phys. **25** 1350018 (2013), `arXiv:0708.1586`.

[7] F. Hélein and J. Kouneiher, Adv. Theor. Math. Phys. **8** 735–777 (2004), `arXiv:math-ph/0401047`.

[8] C. Esposito, *Lectures on Deformation Quantization of Poisson Manifolds* (2012), `arXiv:1207.3287v2 [math-ph]`.

IOP Conf. Series: Journal of Physics: Conf. Series **1030** (2018) 012003 doi:10.1088/1742-6596/1030/1/012003

Searching for Planck scale effects from the free expansion of a Bose–Einstein Condensate

Elías Castellanos[1], **Peter Sloane**[2]

[1 2] Mesoamerican Centre for Theoretical Physics, Universidad Autónoma de Chiapas. Ciudad Universitaria, Carretera Zapata Km. 4, Real del Bosque, 29040, Tuxtla Gutiérrez, Chiapas, México.

E-mail: [1]`ecastellanos@mctp.mx`, [2]`psloane@mctp.mx`

Abstract. The free expansion of a Bose–Einstein condensate is analyzed, assuming that the single particle energy spectrum is modified due to some quantum structure of space time. The condensate's free velocity expansion leads in a natural way to a deformed uncertainty principle. We show that large expansion times are required in order to amplify some possible traces of the Planck scale regime. The analysis presented in this work opens up the opportunity to use Bose–Einstein condensates as tools in *quantum gravity phenomenology*, with a solid experimental basis.

1. Introduction

It is generally accepted that the single particle energy spectrum could be modified due to the possible quantum structure of space time. In ordinary units the non–relativistic form of the aforementioned modified dispersion relation can be expressed as follows [1, 2]

$$\epsilon \simeq mc^2 + \frac{p^2}{2m} + \frac{1}{2M_p}\Big(\xi_1 mcp + \xi_2 p^2 + \xi_3 \frac{p^3}{mc}\Big). \qquad (1)$$

The parameters ξ_1, ξ_2, and ξ_3, are model dependent [1, 2, 3], and should take positive or negative values close to 1 (see Ref. [4] for more details). In fact, the form of the energy dispersion relation (1), was recently constrained by using high precision atom–recoil frequency measurements [1, 2]. In this scenario, bounds for the deformation parameters of order $\xi_1 \sim -1.8 \pm 2.1$ and $-3.8 \times 10^9 < \xi_2 < 1.5 \times 10^9$ were obtained.

It is important to note here that the use of N-body systems as test tools in the search for possible manifestations of Planck scale physics has become a very interesting line of research [4, 5, 6, 7, 8, 9, 10, 11].

These works suggest that some relevant properties associated with Bose–Einstein condensates could be used to obtain representative bounds on the deformation parameters associated with quantum gravity models [9, 12] or for certain areas in the parameter space it is possible to explore the sensitivity of these systems to Planck scale effects [4, 5, 10, 11, 13, 14].

With this in mind, it is well know that when the trapping potential that confines the condensate is turned off, the free velocity expansion of the cloud corresponds to the velocity predicted by Heisenberg's uncertainty principle [15, 16, 17]. This fact is one of several reasons why Bose–Einstein condensates are relevant systems in the analysis and estimation of possible Planck scale effects, since quantum gravity models suggest modifications to this principle [18, 19, 20].

In Refs. [4, 5] we have analyzed the free velocity expansion of a Bose–Einstein condensate, assuming that the energy per particle is modified due to some possible quantum structure of space–time. As a consequence, we were able to deduce a modified uncertainty principle, of the form:

$$\Delta x \Delta p \geq \frac{\hbar}{2}(1+\beta^*) - \alpha^* x + \ldots, \tag{2}$$

where $\alpha^* = 2\xi_1 m^2 c/3\sqrt{\pi}M_p$, M_p is the Planck mass, c is the speed of light, and m is the mass of the particle and $\beta^* \equiv \xi_2 m/M_p$. We must mention that, as far as we know, this linear modification had not previously been reported in the literature [18, 19, 20].

With this at hand, we are able to extend the analysis of the free expansion to more general and realistic scenarios, in which the interaction between particles are present in the system. Moreover, numerical analyses at different time scales in the free expansion were performed, allowing us to determine the time scales in which possible effects related to the quantum structure of space time could be amplified. The analysis of the free expansion of the condensate at different time scales shows that the more promising scenario corresponds to large expansion times, of order of a few seconds, that can be in principle achieved in free fall experiments [17].

2. Modified uncertainty principle and free expansion

Let us calculate the *modified* energy associated with the system. The total energy of the system with contributions of the deformation parameters α and β is given by

$$\begin{aligned} E(\psi) &= \int d\mathbf{r}\left[\frac{\hbar}{2m}|\nabla\psi(\mathbf{r})|^2 + V(\mathbf{r})|\psi(\mathbf{r})|^2 + \frac{U_0}{2}|\psi(\mathbf{r})|^4 \right. \\ &+ \left. \hbar\alpha|\psi(\mathbf{r})|\sqrt{|\nabla|^2}|\psi(\mathbf{r})| + \beta\hbar|\nabla\psi(\mathbf{r})|^2\right], \end{aligned} \tag{3}$$

where ψ is the wave function of the condensate or the so–called order parameter, $V(r) = m\omega_0^2 r^2/2$ is the external potential. The term $U_0 = 4\pi\hbar a/m$ models the interatomic potential, a being the s–wave scattering length, i.e. only two–body interactions are taken into account. Additionally, we have assumed that $\xi_3 = 0$.

In order to calculate the energy of the system at any time let us employ, as usual, the following *ansatz* [15]

$$\psi(\mathbf{r}) = \frac{N^{1/2}}{\pi^{3/4}R^{3/2}}\exp(-r^2/2R^2)\exp[i\phi(r)], \tag{4}$$

where N is the corresponding number of particles and $\phi(r)$ is a phase related to particle flows in the system. If the external potential $V(r)$ is turned offat $t = 0$, there is a force that changes R and produces an expansion of the cloud [15]. It is straightforward to obtain the kinetic energy E_F by using the *ansatz* Eq. (4), with the result $E_F = 3\dot{R}^2 Nm/4$. Moreover, assuming that the energy is conserved at any time, we obtain the following energy conservation condition associated with our system

$$\frac{3m\dot{R}^2}{4} + \frac{3\hbar^2}{4mR^2} + \frac{U_0 N}{2(2\pi)^{3/2}R^3} - \alpha\frac{2\hbar}{\sqrt{\pi}R} + \beta\frac{3\hbar^2}{2R^2} = \frac{3\hbar^2}{4mR_0^2} + \frac{U_0 N}{2(2\pi)^{3/2}R_0^3} - \alpha\frac{2\hbar}{\sqrt{\pi}R_0} + \beta\frac{3\hbar^2}{2R_0^2}, \tag{5}$$

where the dot stands for derivative with respect to time and R_0 is the radius of the condensate at time $t = 0$, which is approximately equal to the oscillator length $a_{ho} = (\hbar/m\omega_0)^{1/2}$ in the non–interacting case. Otherwise, when interactions are present, we will assume that the initial radius corresponds to the result for an isotropic trap [15]

$$R_0 = \left(\frac{2}{\pi}\right)^{1/10}\left(\frac{Na}{a_{ho}}\right)^{1/5} a_{ho}. \tag{6}$$

We remark that R is function of time, i.e., R corresponds to the radius at time t, see (5). If we set $\alpha = \beta = 0$ then we recover the usual solution in the non interacting case [15] which is given by

$$R^2(t) = R_0^2 + \left(\frac{\hbar}{mR_0}\right)^2 t^2. \tag{7}$$

Notice that in the usual case, $\alpha = \beta = 0$, $v_0 = \hbar/mR_0$ is defined as the velocity expansion of the condensate, corresponding to the velocity predicted by Heisenberg's uncertainty principle for a particle confined within a distance R_0 [15]. Thus, in the usual case $\alpha = \beta = 0$, the width of the cloud at time t can be written in its usual form $R^2(t) = R_0^2 + (v_0 t)^2$.

It is interesting to note that when interactions are neglected we are able to obtain an analytical solution for Eq. (5) when $R >> R_0$ together with $\alpha \ll 1$ and $\beta \ll 1$. In such a scenario we obtain

$$R^2_{\alpha,\beta}(t) = R_0^2 + \left[\frac{\hbar}{m^2R_0^2}\left(1 + 2m\beta\right)^2 - \alpha\frac{8}{3\sqrt{\pi}}\frac{\hbar}{mR_0}\right]t^2. \tag{8}$$

If we set $\beta = 0$, the result obtained in Ref. [4] is recovered. Thus, we may recognize the free velocity expansion as a function of the deformation parameters α and β, which is given by

$$(v_0^{\alpha,\beta})^2 = \frac{\hbar}{m^2R_0^2}\left(1 + 2m\beta\right)^2 - \alpha\frac{8}{3\sqrt{\pi}}\frac{\hbar}{mR_0}. \tag{9}$$

Since the corrections caused by α and β are small the following expansion is justified:

$$(v_0^{\alpha,\beta}) \approx \frac{\hbar}{mR_0}(1 + 2m\beta) - \alpha\frac{4}{3\sqrt{\pi}}. \tag{10}$$

Then, the velocity expansion corresponds to the following deformed Heisenberg uncertainty principle

$$\Delta x \Delta p \geq \frac{\hbar}{2}(1 + \beta^*) - \alpha^* x + \ldots, \tag{11}$$

where we have defined $\beta^* \equiv \xi_2 m/M_p$ and $\alpha^* \equiv \xi_1 2m^2c/3\sqrt{\pi}M_p$ together with $R_0 = x$.

Let us now analyze the velocity expansion of our condensate by solving Eq. (5) numerically at any time and taking into account the interactions among the particles within the system. In doing so, we will use some experimental values of interest related to the condensate, i.e., $N \sim 10^4 - 10^6$ particles, $\omega_0 \sim 10 - 10^6$ Hz, $a \sim 10^{-9}$ m, and $m \sim 10^{-26}$ kg [21]. Together with $M_p \simeq 2.18 \times 10^{-8}$kg, $\hbar \sim 6.623 \times 10^{-34}$Js, and $c \sim 3 \times 10^8$m/s. Finally, we assume that the values for the deformation parameters are those reported in Refs. [1, 2].

There are three cases of interest: a) When α,β and U_0 are different from zero. b) For $\alpha = \beta = 0$ and $U_0 = 4\pi h^2 a/m$ and c) when $\alpha, \beta, U_0 = 0$

In Figure 1 the numerical solutions for the modified velocity for the cases a) and b) are illustrated. The velocity in case c) is the usual result for an ideal gas and so we show only the cases a) and b).

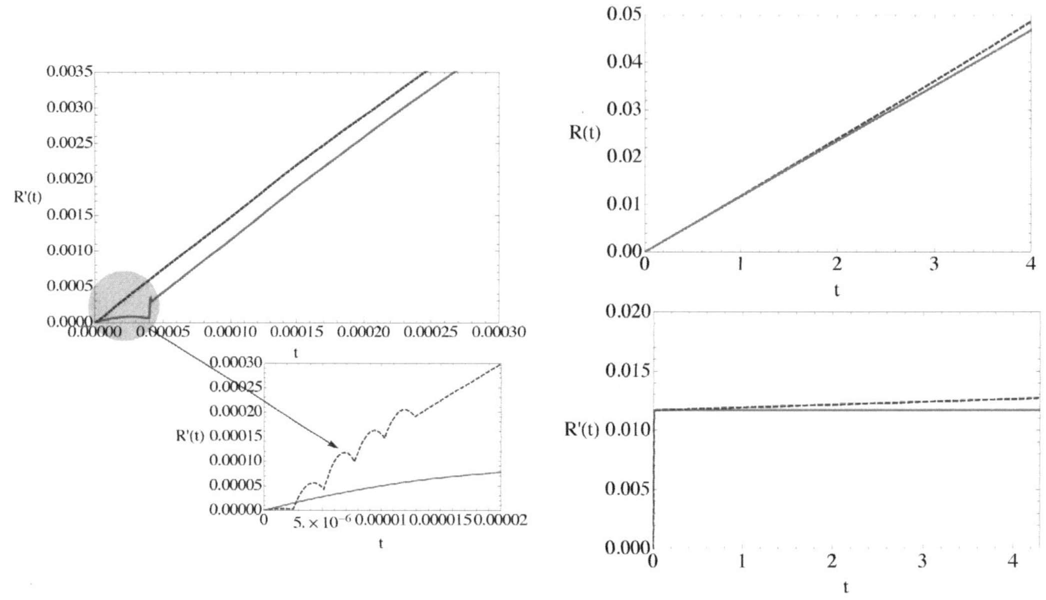

Figure 1. *Left*: Velocity expansion of the condensate at early times for the cases a) and b) are shown in the upper plot. Scenario a) is represented by the solid line and scenario b) is represented by the dashed line. The plot below represents the evolution at early times.
Right: The velocity expansion at large time scales of order $t \sim 4$ sec is plotted above, the plot below represents the free expansion with the same time scale (Figures taken from Ref. [5]).

3. Discussion

Summarizing, we note that at very early expansion times, there is a period in which the velocity seems to be dominated by the deformation parameters. However, we estimate that this short period of expansion of order $t \sim 7.7 \times 10^{-6}$ sec may not be accessible from the experimental point of view, since the results of [16] offer accessibility at the order of milliseconds, i.e., three orders of magnitude bigger than the expansion time obtained here. Conversely, for large expansion times up to $t \sim 4$ sec, there is a region in which the presence of the deformation parameters modified the velocity expansion in a way that may be significant, even when interactions are present.

Concerning the experiment performed in [17], it was proven that for sufficiently large expansion times, the system operates deeply in the linear regime, i.e., almost as in the non-interacting case. In this experiment it was shown by measuring the free velocity expansion at large times that the evolution of the condensate can be independent of interactions during extended free fall experiments. Each of the above scenarios shows that the *modified* free velocity expansion leads to deformations of Heisenberg's uncertainty principle which are around two orders of magnitude smaller than the typical case [5]. This fact could be tested, in principle, in the laboratory, by searching for deviations in the free velocity expansion of the condensate. However, we remark that according to our results, large expansion times are required. This analysis opens a very important branch of research concerning the search for traces of quantum gravity in low energy earth based experiments.

Acknowledgments
E. C. acknowledges MCTP/UNACH for financial support.

References

[1] G. Amelino-Camelia, C. Lämmerzahl, F. Mercati, G. M. Tino, Phys. Rev. Lett. **103** (2009) 171302.
[2] F. Mercati, D. Mazon, G. Amelino-Camelia, J. M. Carmona, J. L. Cortés, J. Induráin, C. Lämmerzahl, G. M. Tino, Class. Quant. Grav. **27** (2010) 215003.
[3] G. Amelino-Camelia, Living Rev. Rel. **16** (2013) 5.
[4] E. Castellanos and J. I. Rivas, Phys. Rev D **91**, 084019 (2015).
[5] Elías Castellanos and Celia Escamilla-Rivera, Mod. Phys. Lett. A **32**,1750007 (2017).
[6] I. Pikovski, M. R. Vanner, M. Aspelmeyer, M. S. Kim, C. Brukner, Nature Phys. **8** (2012) 393-397.
[7] J. D. Bekenstein, Phys. Rev. D, **86** (2012) 124040.
[8] G. Amelino-Camelia, Phys. Rev. Lett. **111** (2013) 101301.
[9] E. Castellanos, G. Chacon-Acosta, Phys. Lett. B **722** (2013) 119-122.
[10] E. Castellanos, EuroPhys. Lett., **103** (2013) 40004.
[11] E. Castellanos, J. I. Rivas and V. Dominguez, EuroPhys. Lett., **106** (2014) 60005.
[12] E. Castellanos, C. Laemmerzahl, Phys. Lett. B **731** (2014) 1–6.
[13] E. Castellanos, C. Laemmerzahl, Mod. Phys. Lett. A **27** (2012) 1250181.
[14] Elías Castellanos, Guillermo Chacon–Acosta, Hector Hugo Hernández–Hernández and Elí Santos, Int. J. Theor. Phys. (2017). https://doi.org/10.1007/s10773-017-3542-0
[15] C. J. Pethick, H. Smith, Bose–Einstein Condensation in Dilute Gases, Cambridge University Press, Cambridge, (20049.
[16] M. R. Andrews, *et. al.* Science 275, 637 (1997).
[17] H. Muntinga et al., Phys. Rev. Lett. 110, 093602 (2013).
[18] A. Kempf, G. Mangano and R. B. Mann, Phys. Rev. D **52** 1108 (1995).
[19] A. F. Ali, S. Das, E. C. Vagenas, Phys. Lett. B **678** 497 (2009).
[20] B. Majumder, S. Sen, Phys. Lett. B **717** 291294 (2012).
[21] F. Dalfovo, S. Giordini, L. Pitaevskii and S. Strangari, Rev. Mod. Phys. 71, 463 (1999).

IOP Conf. Series: Journal of Physics: Conf. Series **1030** (2018) 012004 doi:10.1088/1742-6596/1030/1/012004

Modified relativistic Jüttner-like distribution functions with η-parameter

L. Aragón-Muñoz

Faculdad de Ingeniería, Nuevo Campus Universitario Universidad Autónoma de Chihuahua, Chihuahua 31125, Mexico

E-mail: a257149@uach.mx

G. Chacón-Acosta

Departamento de Matemáticas Aplicadas y Sistemas, Universidad Autónoma Metropolitana-Cuajimalpa, Av. Vasco de Quiroga 4871, México D. F. 05348, Mexico

E-mail: gchacon@correo.cua.uam.mx

Abstract.
In this work we present a family of distribution functions whose modifications are characterized by the parameter η. These parametrization contain all of the different Jüttner-like distributions that have been previously proposed in the literature as an alternative to Jüttner original distribution for relativistic gases in equilibrium. We obtain general expressions for the partition function and for the total energy for some special cases. By analyzing the non relativistic and ultra relativistic limits, we note that only in the latter the parameter η reduces the degrees of freedom in the partition of energy.

1. Introduction

Having a thermodynamic theory in the relativistic regime is important for studying systems in astrophysics [1], cosmology [2] and heavy ions physics [3] where it could be possible to obtain energies large enough to have important relativistic effects. It is also important for black hole thermodynamics [4] and even for quantum gravity theories [5, 6, 7].

The relativistic formulation of thermodynamics has had theoretical complications mainly because the theory of relativity and classical thermodynamics have fundamental differences in their conventional formulations [8]. There are certain efforts to have a Lagrangian [9], Hamiltonian [10], or even a geometrical formulation of thermodynamics [11, 12], but there is still no agreement on it. Since the 1960's there was a controversy about the transformation rules of thermodynamic magnitudes such as temperature [13, 14, 15, 16]. On one hand, relativistic theories emphasizes the transformation properties between inertial frames of the involved variables, whereas the thermal physics approach works on a particular reference system [17, 18], mainly because to date, there are no experiments involving transformation between reference systems, although there are some proposals [19, 20]. For this reason it is usual to choose the microscopic approaches which link the dynamics of the constituents with the macroscopic entities. Statistical physics and kinetic theory in the relativistic regime have been shown to be

IOP Conf. Series: Journal of Physics: Conf. Series **1030** (2018) 012004 doi:10.1088/1742-6596/1030/1/012004

consistent with known results and in recovering the limiting cases of very high and low energies [21, 22].

Within this approach the macroscopical variables are identified with the averages of microscopic dynamical quantities calculated with a particular probability distribution. In the relativistic case the right generalization of the Maxwell-Boltzmann distribution in the momentum space is the so-called Jüttner distribution function [23]. This conclusion was obtained after a debate in which several alternative distributions coming from different approaches were proposed [24, 25, 26, 27, 28, 29], and tested by the numerical experiments done by Cubero et al. [20, 30], Montakhab et al. [31], and also by analytical arguments [32] and covariant studies [33].

However, the proposed alternative distributions have been studied, and in some cases it has been possible to relate them to certain concrete physical situations or probabilistic considerations [34, 35, 36, 37, 38, 39, 40]. Therefore, it is interesting to thoroughly study the different interpretations of the modified Jüttner-like distributions. In this paper we present a family of distributions characterized by an η parameter. For different values of η the different distributions previously proposed are recovered, and we highlight their mathematical and statistical properties. Interesting particular cases and the corresponding statistical moments linked with macroscopic properties are studied.

The structure of the work is as follows: In the next section the family of Jüttner-like η-parametric distributions is established. The partition function is obtained for d dimensions and general η, some particular cases are studied. Section 3 examines the total energy and how it is modified by the η parameter. Finally, a discussion of the possible interpretations of these distributions is given. We have chosen units such that $c = k = 1$, where c is the speed of light and k is the Boltzmann constant.

2. Partition function for η-paremetric Jüttner-like distributions

Let us consider the η-parametric family of distribution functions f_η in momentum space, for a system of relativistic particles with mass m, in d spatial dimensions, defined in the comoving frame [36]:

$$f_\eta(\mathbf{p}) \equiv \frac{1}{Z_\eta}\frac{e^{-\zeta\gamma}}{\gamma^\eta}, \qquad \eta \in \mathbb{Z}, \tag{1}$$

dependence on the momentum appears through the relativistic Lorentz factor γ defined by

$$\gamma^2(\mathbf{p}) = \frac{1}{1 - v^2} = 1 + \frac{p^2}{m^2}, \tag{2}$$

where v is the particle's velocity and $p^2 = p_i\, p^i,\ i = 1, 2, 3, \ldots, d$ is the squared spatial norm of the momentum. In Eq. (1) the parameter $\zeta = m/T$ is the ratio between rest and thermal energies and is often called the *relativistic coolness parameter*. As the energy of the particles is a component of the momentum four-vector $p^\mu = (E, \mathbf{p}) = (m\,\gamma,\, m\,\gamma\,\mathbf{v})$, its magnitude must be a scalar, this introduces the so-called mass-shell relation

$$E^2 = p^2 + m^2. \tag{3}$$

The participation function Z_η is a normalization constant subject to conservation of particles, i.e., the integral of f_η over the entire momentum space must be the density of particles n,

$$n \equiv \int d^d p\, f_\eta = \frac{1}{Z_\eta}\int d^d p\, \frac{e^{-\zeta\gamma}}{\gamma^\eta}. \tag{4}$$

IOP Conf. Series: Journal of Physics: Conf. Series **1030** (2018) 012004 doi:10.1088/1742-6596/1030/1/012004

In general the particle density is a component of the 4-vector particle flow, however, the study of the properties under corresponding transformations of the objects coming from the distributions (1) will be seen elsewhere. In the case $\eta = 0$ of Eq. (4) the well-known Jüttner distribution is recovered, it is proportional to a modified Bessel function of second kind as usual [21, 30, 33].

Let us then determine the corresponding normalization constant Z_η, for cases when $\eta > 0$. To do so, we write the Eq. (4) in spherical coordinates in d dimensions, as f_η has no angular dependence then the integral is written in terms of the d-dimensional norm of the momentum p

$$n = \frac{2\,\pi^{\frac{d}{2}}}{\Gamma\left(\frac{d}{2}\right) Z_\eta} \int_0^{+\infty} dp\, p^{d-1}\, \gamma^{-\eta}\, e^{-\zeta\,\gamma} = \frac{2\,m^d\,\pi^{\frac{d}{2}}}{\Gamma\left(\frac{d}{2}\right) Z_\eta} \int_1^{+\infty} d\gamma\, \gamma^{1-\eta}\, \left(\gamma^2 - 1\right)^{\frac{d-2}{2}}\, e^{-\zeta\,\gamma}, \tag{5}$$

where $2\,\pi^{\frac{d}{2}}/\Gamma\left(\frac{d}{2}\right)$ is the solid angle of the unit sphere in the d-dimensional momentum space in and in the last equality the p and γ variables were exchanged using the mass-shell (3). It is convenient to introduce the function $\mathcal{I}_\eta(\zeta; d)$ defined as the following integral

$$\mathcal{I}_\eta(\zeta; d) = \int_1^{+\infty} d\gamma\, \gamma^{-(\eta-1)}\, \left(\gamma^2 - 1\right)^{\frac{d-2}{2}}\, e^{-\zeta\,\gamma}. \tag{6}$$

In this way the partition function remains in terms of $\mathcal{I}_\eta(\zeta; d)$,

$$Z_\eta(\zeta; d) = 2m^d \frac{\pi^{\frac{d}{2}}}{n\,\Gamma\left(\frac{d}{2}\right)} \mathcal{I}_\eta(\zeta; d). \tag{7}$$

Although in [36] the η-parametric distributions functions for relativistic systems in d-dimensions, were characterized in the context of reference measures for relative entropy, the corresponding normalization factors were not calculated. In [37] the special case for $\eta = 1$ and general d was presented in terms of modified Bessel functions of second kind [41]. This correspondence is obtained because the integral (6) can be reduced as follows

$$\mathcal{I}_1(\zeta; d) = \int_1^{+\infty} d\gamma\, \left(\gamma^2 - 1\right)^{\frac{d-2}{2}}\, e^{-\zeta\,\gamma} = \left(\frac{2}{\zeta}\right)^{\frac{d-1}{2}} \frac{\Gamma\left(\frac{d}{2}\right)}{\sqrt{\pi}} K_{\frac{d-1}{2}}(\zeta). \tag{8}$$

Note that, since the integral inherits some properties of modified Bessel functions then $\mathcal{I}_0(\zeta; d) = -\mathcal{I}_1'(\zeta; d)$, which corresponds to the Jüttner case $\eta = 0$,

$$Z_0(\zeta; d) = \frac{2\,m^d}{n} \left(\frac{2\,\pi}{\zeta}\right)^{\frac{d-1}{2}} K_{\frac{d+1}{2}}(\zeta). \tag{9}$$

On the other hand, the integral (6) in the case of one dimension turns to be a Bickley function or integral of Bessel functions [41, 42, 43], defined as

$$Ki_n(x) \equiv \int_1^{+\infty} d\gamma\, \gamma^{-n}\, \left(\gamma^2 - 1\right)^{-\frac{1}{2}}\, e^{-x\,\gamma}, \tag{10}$$

such that $\mathcal{I}_\eta(\zeta; 1) = Ki_{\eta-1}(\zeta)$. The properties of this function are well studied, from (10) we see that $Ki_0(x) = K_0(x)$. Also, they satisfies

$$Ki_n(x) = \int_x^{\infty} dt\, Ki_{n-1}(t), \tag{11}$$

$$Ki_{-n}(x) = (-1)^n \frac{d^n}{dx^n} Ki_0(x). \tag{12}$$

With these expresssions is possible to calculate (6) for specific values of η, particularly $\mathcal{I}_1(\zeta;1) = K_0(\zeta)$, which is the so-called modified Jüttner case [37]. For the Jüttner case we found $\mathcal{I}_0(\zeta;1) - Ki_{-1}(\zeta) - K_1(\zeta)$, by using (12) and the properties of modified Bessel functions of second class. The behavior of the integral for different values η are shown in Fig. 1.

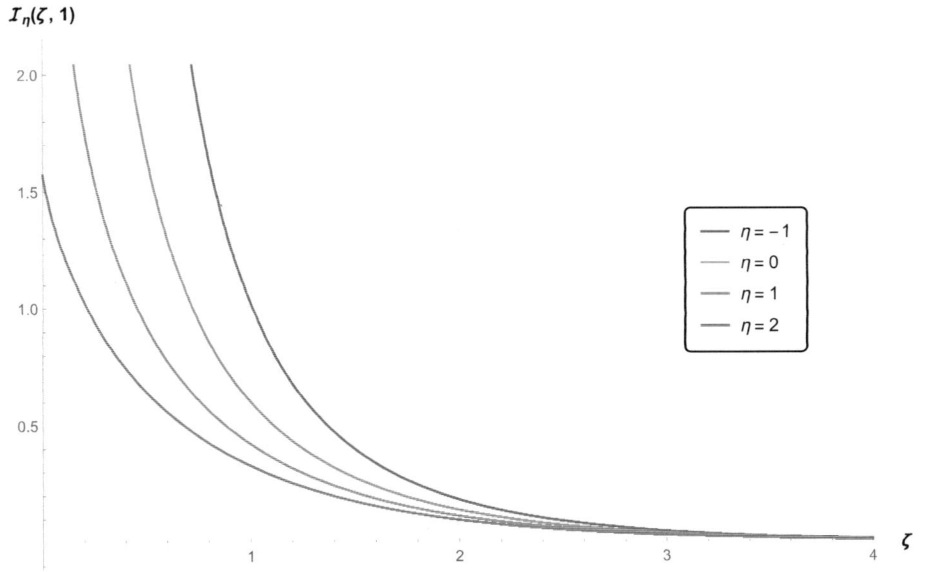

Figure 1. Behavior of the one dimensional case $d = 1$, of the integral $\mathcal{I}_\eta(\zeta; d)$ defined in (6) as function of ζ. Note that as η increases, the integral decreases faster as ζ grows.

These particular cases give us the idea that the general case must necessarily be reducible to Bessel and Bickley functions. To deal with general η, the cases of even ($d = 2m+2$) an odd ($d = 2m+1$) dimensions were considered separately, and the expressions $(\gamma^2-1)^m$ were expanded. In the odd dimensional case $\mathcal{I}_\eta(\zeta; d)$ can be written as a series of Bickley functions, while for even dimension is a sum of E_n exponential integral functions defined as

$$E_n(x) \equiv \int_1^{+\infty} d\gamma\, \gamma^{-n}\, e^{-x\gamma}. \tag{13}$$

In this way the integral for general η in any dimensions can be summarized as follows

$$\mathcal{I}_\eta(\zeta; d) = \begin{cases} \sum_{j=0}^{m} (-1)^j \binom{m}{j} Ki_{\eta-2(m-j)-1}(\zeta), & m = 0,1,2,\dots \ (d \text{ odd}) \\ \sum_{j=0}^{m} (-1)^j \binom{m}{j} E_{\eta-2(m-j)-1}(\zeta), & m = 0,1,2,\dots \ (d \text{ even}) \end{cases} \tag{14}$$

As far as we know these expressions have only been reported in the literature only for specific cases. Note that the effect of the parameter η is on the index of the corresponding function. As we see shall see in next section this could affect the effective degrees of freedom of the statistical quantities.

For general η we can write partition functions for the first three dimensions

$$Z_\eta(\zeta) = \begin{cases} \dfrac{2m}{n} Ki_{\eta-1}(\zeta), & d = 1 \\[2ex] \dfrac{2\pi m^2}{n} E_{\eta-1}(\zeta), & d = 2 \\[2ex] \dfrac{4\pi m^3}{n} \Big(Ki_{\eta-3}(\zeta) - Ki_{\eta-1}(\zeta)\Big). & d = 3 \end{cases} \tag{15}$$

Is worth noticing that these expressions are reduced to the well known $\eta = 0$ case

$$Z_0(\zeta) = \begin{cases} \dfrac{2m}{n} Ki_{-1}(\zeta) = \dfrac{2m}{n} K_1(\zeta), & d = 1 \\[2ex] \dfrac{2\pi m^2}{n} E_{-1}(\zeta) = \dfrac{2\pi m^2}{n} \dfrac{\zeta+1}{\zeta^2} e^{-\zeta}, & d = 2 \\[2ex] \dfrac{4\pi m^3}{n} \Big(Ki_{-3}(\zeta) - Ki_{-1}(\zeta)\Big) = \dfrac{4\pi m^3}{n} \dfrac{K_2(\zeta)}{\zeta}, & d = 3 \end{cases} \tag{16}$$

where the case $d = 1$ was commented above, the case $d = 2$ Is obtained by deriving $E_0(\zeta)$, and for $d = 3$ the expression (12) and the well known recurrence relation for modified Bessel functions of second class $K_{n+1}(x) = K_{n-1}(x) + 2nK_n(x)/x$, were used.

3. Energy and partition theorem for η-modified distributions

In relativistic fluid theory the energy density $n\mathcal{E}/N$ is a component of the energy-momentum tensor which is defined similarly to the particle density (4). However, we can obtain the energy per particle on the comoving frame $\mathcal{E}/N$ from the well-known expression of statistical mechanics

$$\frac{\mathcal{E}_{\eta,d}}{Nm} = -\frac{\partial}{\partial\zeta} \ln Z_\eta(\zeta; d), \tag{17}$$

The main advantage of this expression is that only involves the partition function (7), that is in terms of the integral $\mathcal{I}_\eta(\zeta; d)$, which implies that the energy per particle can be written in a very simple fashion

$$\frac{\mathcal{E}_{\eta,d}}{Nm} = \frac{\mathcal{I}_{\eta-1}(\zeta; d)}{\mathcal{I}_\eta(\zeta; d)}. \tag{18}$$

For $d = 1, 2, 3$, is possible to obtain the internal energy as follows

$$\mathcal{E}_{\eta,d} = Nm \begin{cases} \dfrac{Ki_{\eta-2}(\zeta)}{Ki_{\eta-1}(\zeta)}, & d = 1 \\[2ex] \dfrac{E_{\eta-2}(\zeta)}{E_{\eta-1}(\zeta)}, & d = 2 \\[2ex] \dfrac{Ki_{\eta-4}(\zeta) - Ki_{\eta-2}(\zeta)}{Ki_{\eta-3}(\zeta) - Ki_{\eta-1}(\zeta)}. & d = 3 \end{cases} \tag{19}$$

Due to the properties of Bickley functions and the exponential integral, it is possible to easily recover the Jüttner case [33, 37]. For $d = 1, \mathcal{E}_{0,1}/Nm = K_2/K_1 - \zeta^{-1}$, for $d = 2$ $\mathcal{E}_{0,2}/Nm = 2\zeta^{-1} + \zeta/(1+\zeta)$ and for $d = 3$, $\mathcal{E}_{0,3}/Nm = K_3/K_2 - \zeta^{-1}$.

For general η, the non relativistic limiting case ($\zeta \gg 1$) is obtained from the asymptotic series of Bickley functions [21] and exponential integrals

$$\mathcal{E}_{\eta,d} \approx Nm \begin{cases} 1 + \dfrac{1}{2\zeta} - \dfrac{16\eta^2 + 8\eta - 15}{64\zeta^2} + \ldots, & d = 1 \\ 1 + \dfrac{1}{\zeta} - \dfrac{\eta - 1}{\zeta^2} + \ldots, & d = 2 \\ 1 + \dfrac{3}{2\zeta} - \dfrac{9\left(16\eta^2 - 56\eta + 45\right)}{64\zeta^2} + \ldots. & d = 3 \end{cases} \tag{20}$$

The general behavior of previous expression is $\mathcal{E}_{\eta,d}/Nm = 1 + d/(2\zeta) - \mathcal{O}(\eta, \zeta^{-2})$. Possible effects of the η parameter would appear up to second order ζ^{-2}, so at the lower order of the non relativistic limit the expressions are the same as the case $\eta = 0$, i.e., the equipartition theorem of energy is recovered.

The general expressions in the ultra relativistic case ($\zeta \ll 1$) are complicated, so we study case by case in (19) for different values of the parameter. At least for $d = 1, 2, 3$, it is possible to give the following expression

$$\frac{\mathcal{E}_{\eta,d}}{Nm} \sim \frac{d - \eta}{\zeta}, \qquad \eta \leq d - 1. \tag{21}$$

It is worth noting that the parameter η modifies the effective degrees of freedom with which temperature contributes to the energy of the ultra relativistic gas for $\eta \neq 0$. This indicates that the effect of the η parameter could be verified at very high energies.

Note that expression (21) is only valid for $\eta \leq d - 1$. For $\eta = d,\ d + 1$, the decaying behavior in ζ is very particular and related to the corresponding series of functions in (19), whereas for $\eta \geq d + 2$ the leading energy is constant in temperature. These cases have no physical sense (at least not clearly) and could impose viable values for the η parameter.

4. Discussion

In the relativistic fluid theory, there have been several approaches to describe the interactions of microscopic constituents, which can be summarized in a set of η-parametric distribution functions, such that each value of η corresponds to a different proposal. Several tests were carried out [20, 30, 31, 32, 33] and it was concluded that the case $\eta = 0$, corresponding to the Jüttner distribution, was the correct description of relativistic gases in equilibrium in the comoving frame. However, modified distributions have not been discarded and may even correspond to interesting physical situations [35, 36, 37, 38, 39, 40]. In this work we study the η-parametric family of distributions. We found a general expression for the partition function and for the energy in terms of the η parameter for any spatial dimension. We studied the mathematical properties of the integral $\mathcal{I}_\eta(\zeta; d)$ which can be reduced, in certain cases, to the modified Bessel functions second class, to the exponential integral functions, and to the Bickley functions. For $\eta = 0$, the known results for Jüttner distribution were recovered as expected. Physically, the effect of the parameter can be distinguished only for high energies, where the degrees of freedom with which the temperature contributes to the total energy are reduced by η.

5. Acknowledgements

Support from CONACyT through grant number CB2011/167563 is acknowledged. Also thank A. L. García-Perciante for her useful comments.

IOP Conf. Series: Journal of Physics: Conf. Series **1030** (2018) 012004 doi:10.1088/1742-6596/1030/1/012004

References

[1] Sunyaev R.A. and Zeldovich Y.B. 1972, *Comments Astrophys. Space Phys.* **4** 173.
[2] Bernstein J. 2004, *Kinetic Theory in the Expanding Universe.* (Papeback Ed., Cambridge University Press).
[3] Van Hees H., Greco V., and Rapp R. 2006, *Phys. Rev. C* **73**, 034913.
[4] Wald R. M. 2001, *Liv. Rev, Rel* **4**, 6.
[5] Rovelli C. 1996, *Phys. Rev. Lett.* **77**, 3288.
[6] Gorji M. A. Nozari K., Vakili B. 2014, *Phys. Lett. B* **735**, 62.
[7] Rangamani M. 2009, *Class. Quantum Grav.* **26**, 224003.
[8] Carathéodory C. 1909 *Math. Ann.* **67**, 355.
[9] Glavatskiy K. S. 2015 *J. Chem. Phys.* **142**, 204106.
[10] Baldiotti M. C., Fresneda R. and Molina C. 2016 *Ann. Phys.* **373**, 245.
[11] Mrugala R. 1978 *Rep. Math. Phys.* **14**, 419.
[12] Gambár K. and Márkus F. 1994 *Phys. Rev. E* **50**, 1227.
[13] Planck M. 1909 *Ann. Phys.* (Leipzig) **26**, 1.
[14] Ott H. 1963 *Z. Physik* **175** 70.
[15] Landsberg P. T. 1967 *Nature* **214**, 903.
[16] Landsberg P. T., Matsas G. E. A. 2004 *Phys. A* **340** 92.
[17] Tolman R. C. 1938 *The Principles of Statistical Mechanics.* (Clarendon Press, Oxford).
[18] Veitsman É. 2007 *J. Exp. Theor. Phys.* **105** 1057.
[19] Landsberg P. T., Matsas G. E. A. 1996 *Phys. Lett. A* **223** 401.
[20] Dunkel J., Hänggi P. and Hilbert S. 2009 *Nature Physics* **5**, 741.
[21] Cercignani C. and Kremer G. M. 2002. *The relativistic Boltzmann equation: theory and applications.* (Birkhäuser Verlag, Basel).
[22] Hakim R. 2011 *Introduction to Relativistic Statistical Mechanics: Classical and Quantum.* (World Scientific, Singapore).
[23] Jüttner F. 1911 *Ann. Phys. (Leipzig)* **34**, 856.
[24] Horwitz L. P., Shashoua S. and Schieve W. C. 1989 *Physica A* **161**, 300.
[25] Kaniadakis G. 2002 *Phys. Rev. E* **66**, 056125.
[26] Silva R. and Lima J. A. S. 2005 *Phys. Rev. E* **72**.
[27] Schieve W. C. 2005 *Found. of Physics* **35**, 1359.
[28] Lehmann E. 2006 *J. Math. Phys.* **47**, 023303.
[29] Dunkel J. and Hänggi P. 2007 *Physica A* **374**, 559.
[30] Cubero D., Casado-Pascual J., Dunkel J., Talkner P. and Hänggi P. 2007 *Phys. Rev. Lett.* **99**, 170601.
[31] Montakhab A., Ghodrat M. and Barati M. 2009 *Phys. Rev. E* **79**, 031124.
[32] Debbasch F. 2008 *Physica A* **387**, 2443.
[33] Chacón-Acosta G., Dagdug L., and Morales-Técotl H. A. 2010 *Phys. Rev. E* **81**, 021126.
[34] Dunkel J. and Hänggi, P. 2005 *Phys. Rev. E* **71** 016124.
[35] Fa K. S. 2006 *Braz. J. Phys.* **36**, 777
[36] Dunkel J., Talkner P. and Hänggi P. 2007 *New J. Phys.* **9**, 144.
[37] Cubero D. and Dunkel J. 2009 *EuroPhys. Lett.* **87**, 30005.
[38] Peano F., Marti M., Silva L. O. and Coppa G. 2009 *Phys. Rev. E* **79**, 025701(R).
[39] Dunkel J., Hänggi P. and Weber S. 2009. *Phys. Rev. E* **79**, 010101.
[40] Ghodrat M. and Montakhab A. 2010 *Phys. Rev. E* **82**, 011110.
[41] Abramowitz, M. and Stegun, I. 1964 *Handbook of Mathematical Functions With Formulas, Graphs, and Mathematical Tables.* (National Bureau of Standards Applied Mathematics Series, 55).
[42] Bickley W. G. 1935 *Philos. Mag.* **20**, 322.
[43] Olver F. W. J. et al. 2010 *NIST Handbook of Mathematical Functions.* (NIST-Cambridge University Press, New York).

IOP Conf. Series: Journal of Physics: Conf. Series **1030** (2018) 012005 doi:10.1088/1742-6596/1030/1/012005

Post-Newtonian approximation of gravitational waves from the inspiral phase

Javier M. Antelis

Tecnológico de Monterrey en Guadalajara, Av. Gral. Ramón Corona 2514, Col Nuevo México, C.P. 45201, Zapopan, Jalisco, México

E-mail: mauricio.antelis@itesm.mx

Jaime M. Hernández

Departamento de Ciencias Exactas y Naturales, Centro Universitario de los Valles, Universidad de Guadalajara, Carretera Guadalajara-Ameca km 45.5, C.P. 46600, Ameca, Jalisco, México

E-mail: quijotecuantico@hotmail.com

Claudia Moreno

Departamento de Física, Centro Universitario de Ciencias Exactas e Ingenierías, Universidad de Guadalajara, Av. Revolución 1500, Colonia Olímpica C.P. 44430, Guadalajara, Jalisco, México

E-mail: claudia.moreno@cucei.udg.mx

Abstract. We review the post-Newtonian (PN) expansion for the relativistic correction to the motion of a binary system of black holes. Using the Newtonian formalism, we derive the orbital equations for the 2nd order PN corrections, considering small velocities and weak fields. From a wave equation obtained by the post-Newtonian approximation, we find the h_+ and $h_\times$ polarization of the Gravitational Waves (GW) which could be detected by Laser Interferometer Gravitational Observatory (LIGO).

1. Introduction

A new window to explore our universe will be possible because of the GW discovery [1]. GW were predicted by Einstein in 1916 [2, 3], he described them as *ripples* in the fabric of space-time caused by violent and energetic processes in the Universe such as binary black hole collisions. On February 2016 the first detection of GW was announced. The detection of the binary black hole systems GW150914, GW151226 and GW170104 [4]-[6] was possible due to three fundamental scientific and technological achievements. First, the solution of Einstein's equations through the PN approximation to the relativistic two-body problem and numerical relativity have allowed to obtain accurate GW waveforms during the coalescence of binary BH. Second, the Laser Interferometer Gravitational Observatory (LIGO) [7] detector, which is an extremely sensitive instrument that is able to measure the very tiny GW signals generated by distant astrophysical systems. Third, the data analysis algorithms, which aim to pull out GW signals from the measurements provided by LIGO.

A particular type of GW are produced by orbiting pairs of massive and dense objects, in circular orbits, in this process, the system evolves in three different phases, inspiral, merger and ringdown. In this work we stay in the inspiral phase describing the analytic PN expansion to find the first relativistic corrections to the equations of motion of a binary black holes system. The expansion assumes velocities to be small $v \ll 1$ and a weak field Φ. Using Newtonian approximation, we derive the analytic expression for the second order correction 2PN [8]. The 2PN approximation was useful to construct GW templates for binary systems which were injected in the S5 LIGO run [9, 10].

The present lecture notes are partially based on our paper [11], for a better description of a GW detection, we recommend review it. In section (II) we described the Newtonian approximation formalism. In section (III) we review the second order PN approximation to obtain the gravitational wave $h(t)$, in section (IV) we give some conclusions.

2. Gravitational Theory

2.1. General Relativity

The theory of General Relativity describes space and time as part of the same entity, their representation in any coordinate system is expressed by the line element as

$$ds^2 = g_{\mu\nu} dx^\mu dx^\nu, \tag{1}$$

where $g_{\mu\nu}$[1] is the metric tensor. The representation of the geometry in the Newtonian spacetime is expressed by

$$ds^2 = (1 + 2\Phi)dt^2 - (1 - 2\Phi)\delta_{ij} dx^i dx^j, \tag{2}$$

where Φ is the Newtonian gravitational potential of classical Newtonian gravity.

In the General Relativity theory, the geometry of the spacetime, produced by the matter and energy is described by the *Einstein field equations*

$$G_{\mu\nu} = R_{\mu\nu} - \frac{1}{2} g_{\mu\nu} \mathrm{R} = \frac{8\,\pi\,\mathrm{G}}{c^4} T_{\mu\nu}, \tag{3}$$

where $G_{\mu\nu}$ is the Einstein tensor, $R_{\mu\nu}$ and R are the Ricci tensor and scalar respectively, $T_{\mu\nu}$ is the stress-energy density tensor and G is the Newton's constant of gravitation.

2.2. Gravitational Radiation

If the spacetime is flat with a weak perturbation on it, we can describe it as

$$g_{\mu\nu} = \eta_{\mu\nu} + h_{\mu\nu}, \quad |h_{\mu\nu}| \ll 1, \tag{4}$$

where $\eta_{\mu\nu}$ is the flat Minkowski metric and $h_{\mu\nu}$ is a perturbation to the metric. This $g_{\mu\nu}$ is replaced in equation (3) and the linearized Einstein equations are obtained, which must be invariant under coordinate transformation (gauge transformation) [12]

$$\overline{h}_{\mu\nu} = h_{\mu\nu} + 2\partial_{(\mu}\xi_{\nu)}, \tag{5}$$

where ξ_ν is so small that terms of order $O(h_{\mu\nu}(\partial_\mu \xi^\nu))$ and higher can be neglected of the perturbation. So, the linearized Einstein field equations become

$$\Box \overline{h}_{\mu\nu} + g_{\mu\nu}\partial^\sigma \partial^\rho \overline{h}_{\rho\sigma} - \partial^\rho \partial_\nu \overline{h}_{\mu\rho} - \partial^\rho \partial_\mu \overline{h}_{\nu\rho} = -\frac{16\pi\mathrm{G}}{c^4} T_{\mu\nu}, \tag{6}$$

[1] μ, ν are index which run from 0 to 3 and i, j run from 1 to 3.

where $\Box \equiv \eta_{\mu\nu}\partial_\mu\partial^\mu$ is the D'Alambertian operator[2]. To further simplify this expression, it is convenient to use harmonic gauge such that $\partial^\nu \bar{h}_{\mu\nu} = 0$, which gives

$$\Box\bar{h}_{\mu\nu} = -\frac{16\pi \mathrm{G}}{c^4} T_{\mu\nu}. \tag{7}$$

If we make a measurement very far from the object, the energy-momentum tensor becomes $T_{\mu\nu} = 0$. Substituting the D'Alambertian operator, we obtain the wave equation in vacuum

$$\left(-\frac{1}{c^2}\frac{\partial}{\partial t^2} + \nabla^2\right)\bar{h}_{\mu\nu} = 0, \tag{8}$$

where $\bar{h}_{\mu\nu}$ is the perturbation in the spacetime geometry whose solution is

$$\bar{h}_{\mu\nu} = C_{\mu\nu}\exp(i\kappa_\lambda x^\lambda), \tag{9}$$

where $C_{\mu\nu}$ is a symmetric tensor, transverse-traceless and constant tensor with $C_{0\mu} = C_{\mu 0} = 0$ which contains the polarization of the wave and is called polarization tensor. The spatial components of the GW seen by an observer at a point far from a source, where the source is located at the origin in the xy plane and wave propagates along the z-direction are

$$h_{ab}^{TT} = \begin{pmatrix} h_+ & h_\times \\ h_\times & -h_+ \end{pmatrix}, \tag{10}$$

where h_+ and $h_\times$ are the *plus* and *cross* independent polarizations. Here TT represents Transverse-Traceless tensor because $h_i^i = 0$ and $\kappa^\mu C_{\mu\nu} = 0$, respectively. This shows that the Einstein field equations lead to an oscillatory activity of the spacetime geometry that travel at the speed of light [12].

If we consider $T_{\mu\nu} \neq 0$ into Eq. (7), the solution to the equation becomes

$$\bar{h}_{\mu\nu} = -\frac{16\pi G}{c^4}\int d^4x' G(x-x')T_{\mu\nu}(x'), \qquad G(x-x') = -\frac{\delta(t_{ret}-t')}{4\pi|x-x'|}, \tag{11}$$

where $t_{ret} = t - \frac{1}{c}|x-x'|$ is the retarded time where the GW from source point $\vec{x}'$ needs a certain time to arrive at the observer at position $\vec{x}$, and $G(x-x')$ is the Green function for the 3D wave equation.
When $|\vec{x}| = r$ is greater than the distance of the source d, we can make the expansion $|x-x'| = r - \vec{x}'\cdot\vec{n} + ...$ So the solution to the wave equation is [13]

$$\bar{h}_{\mu\nu} = \frac{4G}{r}\int d^3x'\, T_{\mu\nu}\left(t - r + \vec{x}'\cdot\vec{n}, \vec{x}'\right). \tag{12}$$

Applying Fourier transformation to the source we find

$$T_{\mu\nu}\left(t - r + \vec{x}'\cdot\vec{n}, \vec{x}'\right) = \int \frac{d^4k}{(2\pi)^4}\tilde{T}_{kl}(\omega,\vec{k})\, e^{i\omega(t-r+\vec{x}'\cdot\vec{n})+i\vec{k}\cdot\vec{x}'}. \tag{13}$$

The frequency of the motion inside the source is $\omega_s \sim v/d$. To integrate the Eq. (12) we can expand the exponent as

$$e^{i\omega(t-r+\vec{x}'\cdot\vec{n})} = e^{-i\omega(t-r)}\left[1 - i\omega x'^i n^i + \frac{1}{2}(-i\omega)^2 x'^i x'^j n^i n^j + ...\right], \tag{14}$$

[2] This operator is described by temporal and spatial partial derivatives $\frac{-1}{c^2}\frac{\partial}{\partial t^2} + \nabla^2$.

with this expansion into (13) we can integrate and obtain multipolar expansion, where the quadrupole moment is representative of gravitational waves in Eq. (12),

$$h_{ij} = \frac{2}{r}\frac{d^2M(t)^{ij}}{dt^2}, \tag{15}$$

where $M(t)^{ij}$ is the second mass momenta. As we can see in this expression we have just spatial components in two dimentions, this is because the velocities are small ($v \ll c$) the metric changes slowly, so we can neglect the time derivative of the metric such as the temporal components of h_{00}. In terms of the polarization tensors we have

$$h_x = \frac{1}{r}\frac{G}{c^4}(\ddot{M}_{11} - \ddot{M}_{22}), \qquad h_+ = \frac{2}{r}\frac{G}{c^4}\ddot{M}_{12}. \tag{16}$$

2.2.1. Newtonian approximation for a binary system Coalescence of two black holes is the more intensive source of GW generations. The Newtonian order approximation for a binary inspiral signal is represented by two masses, m_1 and m_2, separated by a distance R (assuming a circular orbit) and orbiting the reduced mass of the system $\mu = \frac{m_1 m_2}{m_1+m_2}$. An orbital plane intersects the y-axis, making an angle ι with the z-axis [11]. Parametrizing the position for a circular orbit $x_1 = R\frac{\mu}{m_1}\hat{e}(t)$ and $x_2 = R\frac{\mu}{m_2}\hat{e}(t)$ where $\hat{e}(t) = [\cos(\omega t), \sin(\omega t)\cos(\iota), \sin(\omega t)\sin(\iota)]$ [13]. The componentes of the spatial tensor in Eq. (16) are

$$M_{11} = R^2\mu\cos^2(\omega t), \quad M_{22} = R^2\mu\sin^2(\omega t)\cos^2(\iota), \quad M_{12} = R^2\mu\cos(\omega t)\sin(\omega t)\cos^2(\iota). \tag{17}$$

Introducing the spatial tensor in the polarization tensor, we obtain

$$h_+(t) = \frac{4}{r}\left(\frac{G\mathcal{M}}{c^2}\right)^{5/3}\left(\frac{\pi f(t)}{c}\right)^{2/3}\frac{1+\cos^2\iota}{2}\cos\phi, \tag{18}$$

$$h_\times(t) = \frac{4}{r}\left(\frac{G\mathcal{M}}{c^2}\right)^{5/3}\left(\frac{\pi f(t)}{c}\right)^{2/3}\cos\iota\sin\phi, \tag{19}$$

where $\mathcal{M} = \mu^{3/5}M^{2/5}$. If we also define the characteristic radius $R_c = \frac{2G\mathcal{M}_c}{c^2}$ and wavelength $\lambda_c = c/f_{gw}$, where f_{gw} is the frecuency of the GW, evaluating at the retarded time, with an arbitrary phase factor, we get

$$h_+(t) = \mathcal{A}\frac{1+\cos^2\iota}{2}\cos\left(2\pi f_{gw}t_{ret} + 2\phi\right), \tag{20}$$

$$h_\times(t) = \mathcal{A}\cos\iota\sin\left(2\pi f_{gw}t_{ret} + 2\phi\right), \tag{21}$$

where the phase and frequency are

$$\phi(t) = -\frac{2c^3}{5GM}\left[\Theta(t)\right]^{-3/8}(t_c - t) + \phi_c, \tag{22}$$

$$f_{gw}(t) = \frac{c^3}{8\pi GM}\left[\Theta(t)\right]^{-3/8}. \tag{23}$$

2.2.2. Post-Newtonian Approximation The Newtonian approximation has not sufficient accuracy to model the quasi-stationary circular orbits in a binary system, to increase accuracy we need the post-Newtonian theory, which adds corrections in the form of a power series in ϕ and f (expanded around the dimensionless parameter $x_{par} = \left(\frac{Gm\omega_s}{c^3}\right)^{2/3}$). The phase and frequency time-evolution of the GW in the second order post-Newtonian (2PN) approximation of general relativity are [14]

$$\phi_{gw}(t) = \phi_0 - \frac{1}{\eta}\left[\Theta(t)^{\frac{5}{8}} + \left(\frac{3715}{8064} + \frac{55}{96}\eta\right)\Theta(t)^{\frac{3}{8}} - \frac{3\pi}{4}\Theta(t)^{\frac{1}{4}}\left(\frac{9\,275\,495}{14\,450\,688} + \frac{284\,875}{258\,048}\eta + \frac{1855}{2048}\eta^2\right)\Theta(t)^{\frac{1}{8}}\right],$$

$$\begin{aligned} f_{gw}(t) &= \frac{c^3}{8\pi GM}\left[\Theta(t)^{-\frac{3}{8}} + \left(\frac{743}{2688} + \frac{11}{32}\eta\right)\Theta(t)^{-\frac{5}{8}} - \frac{3\pi}{10}\Theta(t)^{-\frac{3}{4}}\right. \\ &+ \left.\left(\frac{1855099}{14450688} + \frac{56975}{258048}\eta + \frac{371}{2048}\eta^2\right)\Theta(t)^{-\frac{7}{8}}\right], \end{aligned}$$

where $\Theta = \left(\frac{1}{4x_{dim}}\right)^4$ is a dimensionless time variable defined as $\Theta(t) := \frac{\nu c^3}{5Gm}(t_c - t)$ and $\eta = \frac{m_1 m_2}{(m_1+m_2)^2}$ is the symmetric mass ratio, and t_c is the time at which the coalescence takes place.

Note that these expressions are extensions for the phase and frequency of the GW in the classical Newtonian order solution presented in equations (22) and (23). It is important to mention that to discriminate the signal of a coalesce binary from noise detectors as LIGO, we need the corrections from the post-Newtonian approximation at higher PN orders. In this work just obtain the 2PN approximation which had been used in [11].

3. Gravitational wave strain signal: $h(t)$

Considering the reference frame of an interferometer-based detector, the GW strain produced by a source located at (r, θ, φ) is a linear superposition of the h_+ and $h_\times$ polarizations of the GW according to the detector's response [16]

$$h(t) = F_+ h_+(t) + F_\times h_\times(t), \tag{24}$$

where $F+$ and $F_\times$ are the so-called beam-pattern functions of the detector depending on two angles giving the direction of the source as seen from the detector and a polarization angle, where

$$\begin{aligned} F_+ &= -\frac{1}{2}(1+\cos^2\theta)\cos 2\varphi\cos 2\psi - \cos\theta\cos 2\varphi\sin 2\psi, \\ F_\times &= +\frac{1}{2}(1+\cos^2\theta)\cos 2\varphi\sin 2\psi - \cos\theta\sin 2\varphi\cos 2\psi. \end{aligned} \tag{25}$$

The angle ψ is the third Euler angle that translates from the detectors frame to the radiation frame. Therefore, the GW strain induced in an interferometer-based detector is

$$h(t) = \frac{A(t)}{\mathcal{D}}\cos\left(2\phi_{gw}(t) - \theta\right), \tag{26}$$

where $A(t) = -(2G\mu/c^4)\left(\pi GM f_{gw}(t)\right)^{\frac{2}{3}}$ is the time-dependent quadrupolar amplitude, θ is a phase angle comprising the inclination angle and detector's response $\tan\theta = \frac{F_\times}{F_+}\frac{\cos\iota}{(1+\cos^2\iota)/2}$ and $\mathcal{D}$ is the *effective distance* of the source

$$\mathcal{D} = \frac{r}{\sqrt{F_+^2\left(\frac{1+\cos^2\iota}{2}\right)^2 + F_\times^2\cos^2\iota}}. \tag{27}$$

The effective distance $\mathcal{D}$ is related to real distance r by the geometrical parameters that relates the source orientation to the detector's orientation. Thus, $\mathcal{D}$ is equal to r when the source is *optimally-oriented* (the inclination angle is $\iota = 0$, i.e., the source is located on the z-axis on the observer's line of sight; and the source location is at sky position $\theta = 0$ or $\pi/2$, i.e. above or below the zenith of the detector), and $\mathcal{D}$ is greater than r when the source is *sub-optimally-oriented.*

4. Conclusion

We obtain the second order post-Newtonian approximation for a binary system of two black holes in the inspiral phase at 2PN order. To obtain this approximation we expanded the solution of the Newtonian orbital equations considering negligible velocities $v \ll c$ and weak field Φ. In this approximation we find a wave equation, the solution to this equation are h_+ and $h_\times$ which represents the polarizations of the GW. To discriminating the signal of a coalescing binary from the detector noise, we need relativistic corrections to those solutions in the form of a power series in ϕ and f. Future research in the detection of GW in the inspiral phase with LIGO will need the post-Newtonan approximation to higher orders than 2PN.

Acknowledgements

The authors would like to thank the support of the CONACyT grant No. 271904 and the CONACyT-AEM grants No. 248411 and 262847. CM want to thank PROSNI-UDG 2016 and UDG-CA-813.

Bibliography

[1] Holst M., Sarbach O., Tiglio M., Vallisneri, M. The Emergence of Gravitational Wave Science: 100 Years of Development of Mathematical Theory, Detectors, Numerical Algorithms, and Data Analysis Tools, 2016 Bulletin of the American Mathematical Society
[2] Einstein, A., Zur Algemeinen Relativitätstheorie 778-786, 1915 Preuss. Akad. Wiss. Berlin, Sitzungsber.
[3] Einstein, A. Die Feldgleichungen der Gravitation 844-847, 1916 Preuss. Akad. Wiss. Berlin, Sitzungsber.
[4] Abbott, B. P. et al., Observation of Gravitational Waves from a Binary Black Hole Merger, Phys. Rev. Lett., 2016, 10.1103/PhysRevLett.116.061102
[5] Abbott, B. P. et al., Observation of Gravitational Waves from a 22-Solar-Mass Binary Black Hole Coalescence, Phys. Rev. Lett., 2016, 10.1103/PhysRevLett.116.241103
[6] Abbott, B. P. et al., GW170104: Observation of a 50-Solar-Mass Binary Black Hole Coalescence at Redshift 0.2, Phys. Rev. Lett., 2017, 10.1103/PhysRevLett.118.221101
[7] Abbott, B. P. et al, LIGO: the Laser Interferometer Gravitational-Wave Observatory, Reports on Progress in Physics, 2009 http://stacks.iop.org/0034-4885/72/i=7/a=076901
[8] Hannam, M. Modelling gravitational waves from precessing black-hole binaries: progress, challenges and prospects, General Relativity and Gravitation 2014 DOI 10.1007/s10714-014-1767-2
[9] Vallisneri, M., Kanner, J., Roy W., Weinstein, A. and Branson S. The LIGO open center, Journal of Physics: Conference Series, 2015 http://stacks.iop.org/1742-6596/610/i=1/a=012021
[10] Abadie, J. et al.,Search for gravitational waves from compact binary coalescence in LIGO and Virgo data from S5 and VSR1, Phys. Rev. D, 2010 DOI 10.1103/PhysRevD.82.102001
[11] Antelis, J. M. and Moreno, C. Obtaining gravitational waves from inspiral binary systems using LIGO data, Eur. Phys. J. Plus, 2017 DOI 10.1140/epjp/i2017-11283-5, 10.1140/epjp/i2017-11396-9
[12] Carroll, S. M. Spacetime and geometry: An introduction to general relativity, ed. Addison Wesley, 2004, ISBN 0-8053-8732-3
[13] Maggiore, M. Gravitational Waves: Volume 1: Theory and Experiments, Oxford University Press, 2007
[14] Blanchet, L., Iyer, B. R., Clifford M. W. and Wiseman, A. G. Gravitational waveforms from inspiralling compact binaries to second-post-Newtonian order, Classical and Quantum Gravity, 1996, http://stacks.iop.org/0264-9381/13/i=4/a=002
[15] Abbott B. P. et. al., Prospects for Observing and Localizing Gravitational-Wave Transients with Advanced LIGO and Advanced Virgo, Living Reviews in Relativity , 2016 DOI 10.1007/lrr-2016-1
[16] Schutz, B. F. Networks of gravitational wave detectors and three figures of merit, Classical and Quantum Gravity, 2011 http://stacks.iop.org/0264-9381/28/i=12/a=125023

IOP Conf. Series: Journal of Physics: Conf. Series **1030** (2018) 012006 doi:10.1088/1742-6596/1030/1/012006

Towards a calculation of the halo mass function of a scalar field dark matter

Francisco X. Linares Cedeño[a], **Alma X. González-Morales**[a,b] **and L. Arturo Ureña-López**[a]

[a]Departamento de Física, DCI, Campus León, Universidad de Guanajuato, 37150, León, Guanajuato, México.

[b]Consejo Nacional de Ciencia y Tecnología, Av. Insurgentes Sur 1582. Colonia Crédito Constructor, Del. Benito Juárez C.P. 03940, México D.F. México

Abstract. In this work we are going to build the halo mass function for a scalar field dark matter model. This will be done with the aim of determine the minimum size of halos that our model predicts, in contrast with the LCDM model, which predicts an overabundance of substructures. We find that, for a scalar field mass $m = 10^{-22}$eV and a Sharp-k window function, is possible to reproduce the suppression in the halo formation at small scales.

1. Introduction

The galaxy distribution on the Universe provides information about the formation and evolution of large scale structures. Several astronomical and cosmological surveys like BOSS [1], PLANCK [2], DES [3], have been exploring the content of matter and energy in the Universe for the last years, providing us with useful information to constraint theoretical models with higher precision. The commonly named LCDM model is so far the theoretical model that successfully describe several cosmological observations [2]; nonetheless, it is well known that there are some tension between small scales observations and N-body simulations based on LCDM. As we will see later, LCDM model predicts an overabundance of galactic substructures that have not been observed. An alternative candidate that have been extensively studied in recent literature is the *axion*, which is represented by a scalar field ϕ endowed with a potential energy density $V(\phi)$. This particle was originally proposed to solve the strong CP problem in QCD [4, 5], and also appears in some string theory scenarios [6–11]. Our aim in this work is to show that an axion-like particle as scalar field dark matter (SFDM) is a good candidate to play the role of dark matter in the Universe, which means that this scalar field fits cosmological observations such as CMB anisotropies and Mass Power Spectrum (MPS) as good as CDM, and additionally predicts a cut-off in the sub-haloes production, which can be seen in the Halo Mass Function (HMF). Some reviews of this SFDM model can be found at [12–14].

We will be focused in the calculation of the HMF for a scalar field dark matter. Calculations of the HMF can be computed as has been shown in [15–18]. Particularly in this work we are going to follow the procedure of reference [15] for a scalar field with a quadratic potential

$$V(\phi) = \frac{1}{2}m^2\phi^2 \,, \tag{1}$$

which has been considered in several cosmological scenarios for dark matter models in literature [19–22].

2. Power Spectra for a Scalar Field Dark Matter

The Einstein equations for a scalar field in a Friedmann-Robertson-Walker (FRW) spacetime with null spatial curvature are given by

$$H^2 = \frac{\kappa^2}{3}\left(\sum_j \rho_j + \rho_\phi\right), \dot{H} = -\frac{\kappa^2}{2}\left[\sum_j (\rho_j + p_j) + (\rho_\phi + p_\phi)\right], \tag{2a}$$

$$\dot{\rho}_j = -3H(\rho_j + p_j), \quad \ddot{\phi} = -3H\dot{\phi} - m^2\phi, \tag{2b}$$

where $\kappa^2 = 8\pi G$, a dot denotes derivative with respect to cosmic time t, H is the Hubble parameter and ϕ is the ALP that will be the source of dark matter. The scalar field energy density ρ_ϕ and pressure p_ϕ are obtained from the energy-momentum tensor of the scalar field, and are given by the canonical expressions:

$$\rho_\phi = \frac{1}{2}\dot{\phi}^2 + \frac{1}{2}m^2\phi^2, \quad p_\phi = \frac{1}{2}\dot{\phi}^2 - \frac{1}{2}m^2\phi^2. \tag{3}$$

In order to transform the Klein-Gordon (KG) equation (2b), we define a new set of variables based on [23, 24] as follows

$$\Omega_\phi^{1/2}\sin(\theta/2) \equiv \frac{\kappa\dot{\phi}}{\sqrt{6}H}, \quad \Omega_\phi^{1/2}\cos(\theta/2) \equiv \frac{\kappa m\phi}{\sqrt{6}H}, \quad y_1 \equiv -\frac{2m}{H}. \tag{4}$$

The KG equation can be written as a first order differential equations system as follows:

$$\theta' = -3\sin\theta + y_1, \quad y_1' = \frac{3}{2}(1 + w_{tot})\, y_1, \quad \Omega_\phi' = 3(w_{tot} - w_\phi)\Omega_\phi, \tag{5}$$

where a prime denotes derivative with respect to the number of e-foldings $N \equiv \ln(a/a_i)$, with a the scale factor of the Universe and a_i its initial value. Now, we consider small perturbations around the background values of the FRW line element (in the synchronous gauge) as well as for the scalar field in the following form:

$$ds^2 = -dt^2 + a^2(t)(\delta_{ij} + h_{ij})dx^i dx^j, \quad \phi(x,t) = \phi(t) + \varphi(x,t), \tag{6}$$

where h_{ij} and φ are the metric and scalar field perturbations respectively. Then, the linearized KG equation reads [25–28]:

$$\ddot{\varphi} = -3H\dot{\varphi} - \left(\frac{k^2}{a^2} + \frac{\partial^2 V(\phi)}{\partial\phi^2}\right)\varphi - \frac{1}{2}\dot{\phi}\dot{h}, \tag{7}$$

where h is the trace of scalar metric perturbations, and k is a comoving wavenumber. Although functional dependence of the scalar field perturbation is not explicitly shown, note that Eq. (7) is written for a Fourier mode $\varphi(k,t)$. Following the change of variables of [19, 29], the Eq.(7) is described by the following system of equations

$$\delta_0' = [-3\sin\theta - \omega(1-\cos\theta)]\,\delta_1 + \omega\sin\theta\,\delta_0 - \frac{h'}{2}(1-\cos\theta), \tag{8a}$$

$$\delta_1' = [-3\cos\theta - \omega\sin\theta]\,\delta_1 + \omega(1+\cos\theta)\delta_0 - \frac{h'}{2}\sin\theta, \tag{8b}$$

where $\omega = k^2/k_J^2 = k^2/H^2a^2y_1$. With the solutions of Eqs. (8) is possible to build up cosmological observables such as CMB anisotropies and the MPS, which we show for the axion field and CDM in Figs. 1 and 2, for an axion mass $m = 10^{-20}, 10^{-22}, 10^{-24}, 10^{-26}$eV. The numerical solutions were obtained with an amended version of the CLASS code[1], which is a Boltzmann code to solve the linear perturbations of the cosmological dynamics. Figure 1 shows

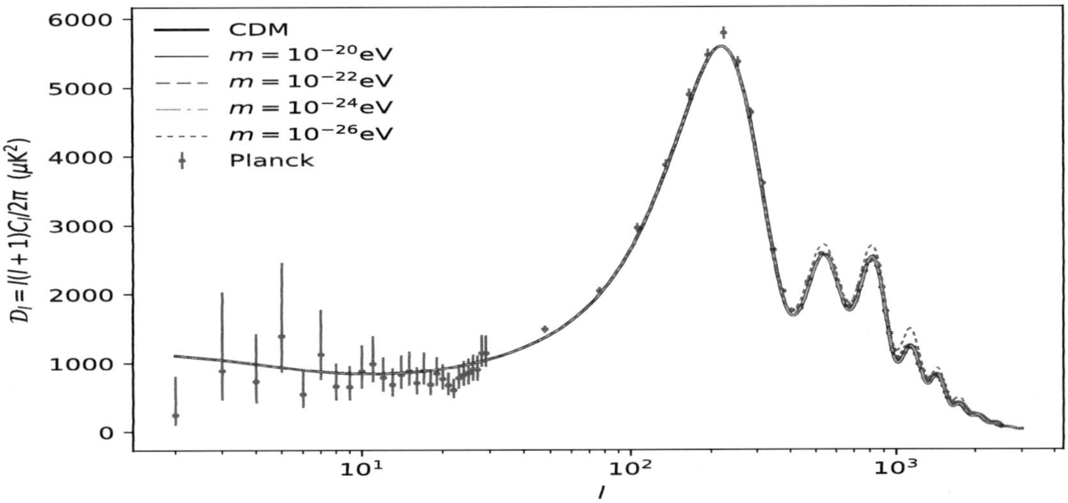

Figure 1: CDM and SFDM Temperature Power Spectra for an axion with masses $m = 10^{-20}, 10^{-22}, 10^{-24}, 10^{-26}$eV. It can be seen that major discrepancies between both models appear at large multipoles for an axion mass $m = 10^{-26}$eV. Purple dots are used to show data from Planck.

a very good agreement between our model and observations from Planck satellite data at least for $m \geq 10^{-24}$eV. On the other hand, Fig. 2 shows the MPS for both CDM and SFDM model. In such figure can be noted that there is a cut-off at large values of k (small scales) which shifts to small wavenumber for smaller masses. In particular, we see that for $m = 10^{-26}$eV the axion is ruled out due to it is not in agreement with the observed MPS at linear regime.

3. Halo Mass Function

Now we are going to calculate the HMF for this dark matter model. Such function encodes the comoving number density of dark matter haloes as function of the halo mass, and it constitutes a representative cosmological probe of dark matter and dark energy. It can be use to constraint the value of the combined parameters σ_8 (power spectrum normalization) and Ω_M (matter density parameter), and also to characterize the dark energy equation of state ω_0 [31–33]. To accomplish this, it is necessary to use a window function, which is a filter function that fixes a cut-off at certain length scale r. Following the recipe of [15] let us proceed as follows: first, we have to define the window function we are going to implement. In this work we will consider the Top-Hat W_{TH}, which is a filter with spherical form in real space, and the Sharp-k W_{SK}, which is defined as a Top-Hat function in Fourier space. The choice of those window functions is based on that the Top-Hat function is useful to work with the LCDM model, while the Sharp-k function does a better job with suppressed power spectra, which is the case for scalar field dark matter. An analysis using those window functions incorporating a mass dependent barrier, and

[1] Such code can be found in https://github.com/lurena-lopez/class.FreeSF

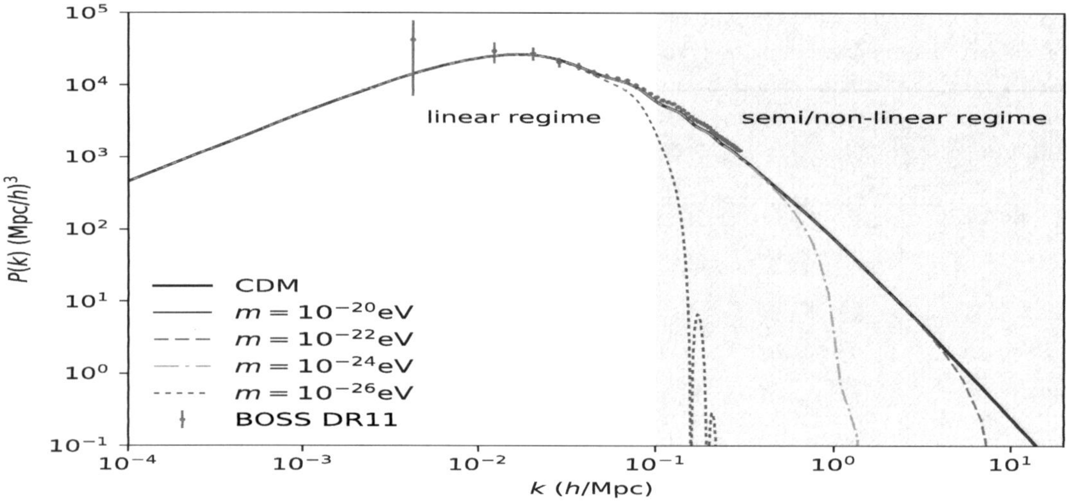

Figure 2: MPS for SFDM with axion mass $m = 10^{-22}$eV. It can be noted that the cut-off at small scales (large wavenumbers) depends on the axion mass m. Data from BOSS [30] are indicated with red dots. See text for more discussion.

also including a comparison with numerical simulations can be found in [19]. Both window function mentioned above are given (in Fourier space) by

$$W_{TH}(kr) = \frac{3}{(kr)^3}\left[\sin(kr) - kr\cos(kr)\right], \quad W_{SK}(kr) = \Theta(1-kr), \tag{9}$$

which encloses a mass given by $M = 4\pi\bar{\rho}r^3/3$ and $\bar{\rho} = 2.775 \times 10^{11}h^2M_{\odot}/\text{Mpc}^3$ is the critical density of the Universe. Now let us define the variance $S(r)$

$$S(r) = \int \frac{d^3\vec{k}}{(2\pi)^3}P(k)W^2(kr). \tag{10}$$

The universal growth factor of perturbations is given by the following expression

$$D(a) = \frac{5\Omega_m}{2}H(a)\int\frac{da}{a^3H^3(a)}, \tag{11}$$

while the peak height of perturbations, the critical overdensity and the function for a spherical collapse model are given respectively by

$$\nu = \frac{\delta_c^2}{S(r)}, \quad \delta_c = \frac{1.686}{D(a)}, \quad f(\nu) = \sqrt{\frac{2\nu}{\pi}}e^{-\nu/2}. \tag{12}$$

Finally, the HMF has the following expression

$$\frac{dn}{d\ln M} = -\frac{1}{2}\frac{\bar{\rho}}{M}f(\nu)\frac{d\ln S}{d\ln M}. \tag{13}$$

The HMF is shown in Fig. 3 as function of halo mass. We note in such figure that the cut-off in the mass power spectrum (Fig. 2) is translated into a suppression of small scale structures.

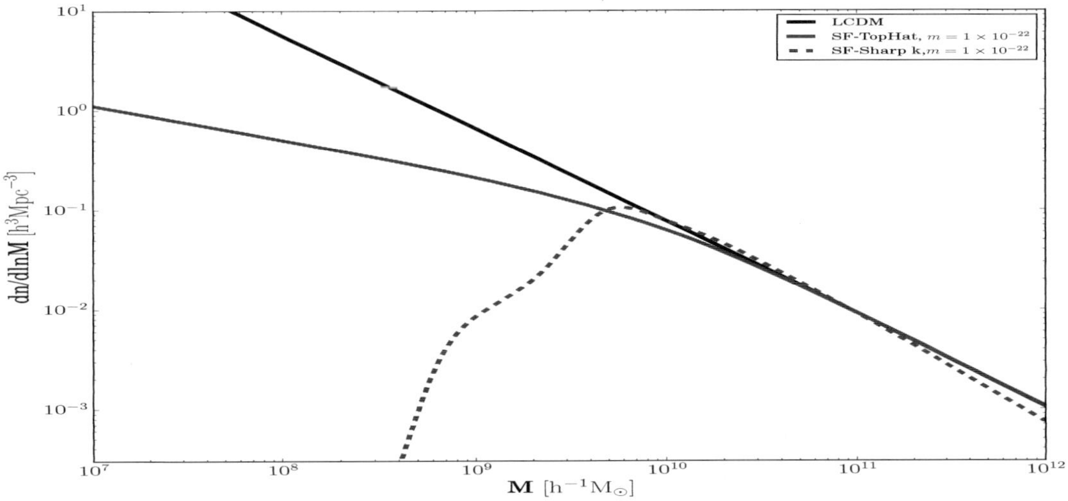

Figure 3: HMF as function of halo mass. The fiducial LCDM model is shown in black line, while the axion case for a mass $m = 10^{-22}$eV is shown for a Top-Hat (blue solid line) and Sharp-k (dotted blue line) window functions. The suppression for the scalar field at small scales is better reproduced by the Sharp-k window function. See text for more discussion.

While numerical simulations are needed to build up the HMF, here we present a forecast for the HMF of our model, particularly for the axion mass $m = 10^{-22}$eV: our results show that for a suppressed power spectrum, the Sharp-k function seems to reproduce better the absence of substructures at small scales. This can be compared with the smaller galactic objects that have been observed so far in the Local Group, the so called dwarf spheroidal galaxies, whose total mass (stellar and DM masses) are of order of $10^7 - 10^8 M_\odot$ [34–36], *i.e.*, the SFDM model predicts a lower bound for halo masses that is consistent with astrophysical observations.

4. Conclusions

In this work we obtain the MPS and the anisotropies of the cosmic microwave background for a dark matter model based on an ultra-light scalar field (axion-like particle). The cosmological implication of such model is evident in the structure formation: for axion with masses $m \leq 10^{-20}$eV there is a shift to small wavenumbers (large scales). This imply that there is a cut-off in the size of galactic structures. This can be seen in Fig. 3, which constitute the main result of this work: while the LCDM model predicts an overabundance of substructures at small masses, the SFDM shows a decrease in the number of halos for low values of halo mass. Particularly, using a Sharp-k window function we observe a suppression for halo masses, this is, the SFDM predicts a lower bound for the size of halo masses smaller than $4 \times 10^8 M_\odot$. It is important to note that this is a crude approximation, in the sense that full cosmological simulations of structure formation including the presence of baryons are needed to quantify in detail the amount of substructures[37, 38]. This is work in progress that we expect to publish elsewhere.

References

[1] Alam S *et al.* (BOSS) 2016 *Submitted to: Mon. Not. Roy. Astron. Soc.* (*Preprint* 1607.03155)
[2] Ade P A R *et al.* (Planck) 2016 *Astron. Astrophys.* **594** A13 (*Preprint* 1502.01589)
[3] Abbott T *et al.* (DES) 2016 *Mon. Not. Roy. Astron. Soc.* **460** 1270–1299 (*Preprint* 1601.00329)
[4] Peccei R D and Quinn H R 1977 *Phys. Rev. Lett.* **38** 1440–1443
[5] Peccei R D and Quinn H R 1977 *Phys. Rev.* **D16** 1791–1797

IOP Conf. Series: Journal of Physics: Conf. Series **1030** (2018) 012006 doi:10.1088/1742-6596/1030/1/012006

[6] Svrcek P and Witten E 2006 *JHEP* **06** 051 (*Preprint* hep-th/0605206)
[7] Witten E 1984 *Phys. Lett.* **B149** 351–356
[8] Ringwald A 2014 *J. Phys. Conf. Ser.* **485** 012013 (*Preprint* 1209.2299)
[9] Cicoli M, Goodsell M and Ringwald A 2012 *JHEP* **10** 146 (*Preprint* 1206.0819)
[10] Cicoli M 2014 *J. Phys. Conf. Ser.* **485** 012064 (*Preprint* 1209.3740)
[11] Ahn Y H 2016 *Phys. Rev.* **D93** 085026 (*Preprint* 1604.01255)
[12] Magaña J and Matos T 2012 *J. Phys. Conf. Ser.* **378** 012012 (*Preprint* 1201.6107)
[13] Marsh D J E 2016 *Phys. Rept.* **643** 1–79 (*Preprint* 1510.07633)
[14] Hui L, Ostriker J P, Tremaine S and Witten E 2017 *Phys. Rev.* **D95** 043541 (*Preprint* 1610.08297)
[15] Schneider A, Smith R E and Reed D 2013 *Mon. Not. Roy. Astron. Soc.* **433** 1573 (*Preprint* 1303.0839)
[16] Schneider A 2015 *Mon. Not. Roy. Astron. Soc.* **451** 3117–3130 (*Preprint* 1412.2133)
[17] Schive H Y, Chiueh T, Broadhurst T and Huang K W 2016 *Astrophys. J.* **818** 89 (*Preprint* 1508.04621)
[18] Du X, Behrens C and Niemeyer J C 2017 *Mon. Not. Roy. Astron. Soc.* **465** 941–951 (*Preprint* 1608.02575)
[19] Ureña López L A and González-Morales A X 2016 *JCAP* **1607** 048 (*Preprint* 1511.08195)
[20] Marsh D J and Ferreira P G 2010 *Phys.Rev.* **D82** 103528 (*Preprint* 1009.3501)
[21] Marsh D J, Tarrant E R, Copeland E J and Ferreira P G 2012 *Phys.Rev.* **D86** 023508 (*Preprint* 1204.3632)
[22] Marsh D J E and Silk J 2013 *Mon.Not.Roy.Astron.Soc.* **437** 2652 (*Preprint* 1307.1705)
[23] Copeland E J, Liddle A R and Wands D 1998 *Phys.Rev.* **D57** 4686–4690 (*Preprint* gr-qc/9711068)
[24] Ureña López L A 2016 *Phys. Rev.* **D94** 063532 (*Preprint* 1512.07142)
[25] Ratra B 1991 *Phys.Rev.* **D44** 352–364
[26] Ferreira P G and Joyce M 1998 *Phys.Rev.* **D58** 023503 (*Preprint* astro-ph/9711102)
[27] Ferreira P G and Joyce M 1997 *Phys.Rev.Lett.* **79** 4740–4743 (*Preprint* astro-ph/9707286)
[28] Perrotta F and Baccigalupi C 1999 *Phys.Rev.* **D59** 123508 (*Preprint* astro-ph/9811156)
[29] Cedeño F X L, González-Morales A X and Ureña López L A 2017 (*Preprint* 1703.10180)
[30] Anderson L *et al.* (BOSS) 2014 *Mon. Not. Roy. Astron. Soc.* **441** 24–62 (*Preprint* 1312.4877)
[31] Vikhlinin A *et al.* 2009 *Astrophys. J.* **692** 1060–1074 (*Preprint* 0812.2720)
[32] Allen S W, Evrard A E and Mantz A B 2011 *Annual Review of Astronomy and Astrophysics* **49** 409–470 (*Preprint* https://doi.org/10.1146/annurev-astro-081710-102514) URL https://doi.org/10.1146/annurev-astro-081710-102514
[33] Murray S, Power C and Robotham A 2013 *Mon. Not. Roy. Astron. Soc.* **434** L61 (*Preprint* 1306.5140)
[34] Sawala T, Guo Q, Scannapieco C, Jenkins A and White S D M 2011 *Mon. Not. Roy. Astron. Soc.* **413** 659 (*Preprint* 1003.0671)
[35] Oh S H, de Blok W J G, Brinks E, Walter F and Kennicutt Jr R C 2011 *Astron. J.* **141** 193 (*Preprint* 1011.0899)
[36] Diez-Tejedor A, Gonzalez-Morales A X and Profumo S 2014 *Phys. Rev.* **D90** 043517 (*Preprint* 1404.1054)
[37] Zhang J, Tsai Y L S, Cheung K and Chu M C 2016 (*Preprint* 1611.00892)
[38] Zhang J, Kuo J L, Liu H, Tsai Y L S, Cheung K and Chu M C 2017 (*Preprint* 1708.04389)

IOP Conf. Series: Journal of Physics: Conf. Series **1030** (2018) 012007 doi:10.1088/1742-6596/1030/1/012007

Rapid cosmological simulations in Scalar Field Dark Matter models

Rebeca Martínez-Carrillo[1] **and L. Arturo Ureña-López**[2]

[1] División de Ciencias e Ingenierías, Campus León, Universidad de Guanajuato, 37150, León, Guanajuato, México

[2] Departamento de Física, DCI, Campus León, Universidad de Guanajuato, 37150, León, Guanajuato, México.

E-mail: r.martinezcarrillo@ugto.mx[1], lurena@ugto.mx[2]

Abstract. The process of structure formation in the Universe can be explained by means of the gravitational instability of the different matter components, in particular that of the so-called dark matter field. In this work, we consider that dark matter can be described by an ultra-light scalar field minimally coupled to gravity, and its gravitational instability in an expanding universe is studied with the help of standard techniques for fluid matter components. Analytical and numerical solutions are shown, for both linear and non-linear perturbations, and compared with those of the standard cold dark matter model. We conclude that non-linear perturbations of the scalar field are indistinguishable from those of cold dark matter as long as the field mass is larger than 10^{-25}eV.

1. Introduction

With a contribution of 25.8% of the total density in the Universe, dark matter (DM) is very important in the context of the structure formation that we observe in our Universe. The most accepted of all is the Λ Cold Dark Matter (ΛCDM) model, which is in very good agreement with several observations at large scales [1]. However, this model presents some problems when compared with observations at small scales, which are summarized as follows:

- The missing satellites problem: Overpopulation of the substructures predicted by the N-body simulations of the ΛCDM model [2–4] is at variance with observations of satellite galaxies of the local group (Milky Way and Andromeda).
- The core-cusp problem. Observations show a shallow density profile at the center of many galaxies, however N-body numerical simulations predict a divergent behavior of the density profile, the called *cusp* density profile [5, 6].
- Too Big, too fail. The Milky Way does have some satellites, as do other galaxies, but when these smaller galaxies are examined they seem to have a lot less DM than suggested by the simulations [7–9].

The problems above motivate the study of the structure formation with alternative models of dark matter. In this paper, we show cosmological simulations in which DM is modeled with an ultra-light scalar field [10–17]. A brief description of the contents is the following. In Sec. 2, we present a brief mathematical background of the dark matter model with a scalar field (details of which can be found in Refs. [16, 17]. In Sec. 3, we describe the numerical codes and parameters to perform cosmological simulations, together with a presentation of the general results. Finally, Sec. 4 is devoted to the general conclusions of this work.

2. Mathematical background

DM is modeled by a scalar field ϕ endowed with a quadratic scalar field potential, with the Lagrangian density

$$\mathcal{L} = -\frac{1}{2}\partial^\mu \phi \partial_\mu \phi - \frac{1}{2} m^2 \phi^2 \,, \tag{1}$$

where m is the boson mass. For the background evolution, we consider the line element of a flat, homogeneous and isotropic universe in the form: $ds^2 = -dt^2 + a^2(t)\delta_{ij}dx^i dx^j$, where $a(t)$ is the scale factor. The equations of motion are then

$$H^2 = \frac{8\pi G}{3}\left[\sum_j \bar{\rho}_j + \frac{1}{2}\dot{\phi}^2 + \frac{1}{2}m^2\phi^2\right] , \tag{2}$$

$$\ddot{\phi} = -3H\dot{\phi} - m^2\phi \,, \tag{3}$$

with $H = \dot{a}/a$ being the Hubble parameter and $\sum_j \bar{\rho}_j$ is the collective contribution of other matter components. Also, a dot denotes derivative with respect to the cosmic time t.

Likewise, we consider linear perturbations to the background evolution, with a line element written in the synchronous gauge as $ds^2 = -dt^2 + a^2(t)(\delta_{ij} + h_{ij})dx^i dx^j$, where h_{ij} are the spatial perturbations of the metric (more details in Ref. [18]). For the scalar field, we write it as a background part plus a linear perturbation, $\phi(x,t) = \phi(t) + \varphi(x,t)$, and then the equation of motion of the scalar field perturbation φ in Fourier space for a given wavelength k is

$$\ddot{\varphi} = -3H\dot{\varphi} - \left(\frac{k^2}{a^2} + m^2\right)\varphi - \frac{1}{2}\dot{\phi}\dot{h} \,. \tag{4}$$

A full solution of the system of equations (2), (3) and (4) can only be obtained by numerical means, in which the scalar field perturbations can also be coupled to other matter perturbations for a correct calculation of different observables. In the present case, we use the amended version of the CLASS code that was presented in Ref. [16], from which we computed the different instances of the mass power spectrum (MPS) shown in Fig. 1. The curves in color represent the case of the scalar field with different boson masses in the range $m = 10^{-22} - 10^{-25}$ eV.

Also shown is the standard cosmological result of ΛCDM, and then we can notice that in comparison the scalar field cases present a sharp cut-off, which means the suppression of power at small length scales. The position of the cut-off, and then the scales that are suppressed, depends upon the boson mass, and in general occurs for larger scales for lighter boson masses.

3. Cosmological simulations and general results

In order to get the non-linear evolution of the cosmological perturbations, we require to evolve the MPS shown in Fig. 1 by means of an N-body code. N-body simulations are simulations of a dynamical system of particles under the influence of forces, like gravity. In general, numerical codes for N-body simulations solve the evolution of the particles in terms of the equation of motion [19]

$$\frac{d^2\mathbf{\Psi}}{d\eta^2} + H(\eta)\frac{d\mathbf{\Psi}}{d\eta} + \nabla\Phi = 0 \,, \tag{5}$$

where η is the conformal time, and the displacement vector $\mathbf{\Psi}$ is defined as

$$\mathbf{x}(\eta) = \mathbf{q} + \mathbf{\Psi}(\mathbf{q}, \eta), \tag{6}$$

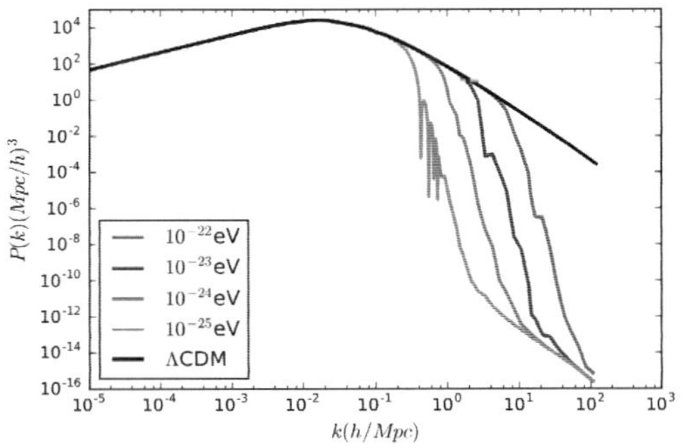

Figure 1. MPS with a scalar field as DM considering different values of the boson mass m, see Eq. (1). The most noticeable feature, in comparison with the case of ΛCDM, is the appearance of sharp cut-off for large wavelengths k. This cut-off depends on the mass parameter and could in principle be used to put constraints on the value of m, see for instance Ref. [16, 17] and references therein.

with $\mathbf{x}$ and $\mathbf{q}$ the Eulerian and Lagrangian coordinates, respectively. The gravitational potential Φ in Eq. (5) obeys the Poisson equation,

$$\nabla^2\Phi = \frac{3}{2}\Omega_m H\delta(\mathbf{x}), \tag{7}$$

where Ω_m is the matter density parameter (obtained from the background evolution) and $\delta(\mathbf{x})$ is the density contrast[1], which is calculated from the expression

$$\delta(\mathbf{x}) = \frac{1}{\mathrm{Det}(\delta_{ij} + \partial\Psi_i/\partial q_j)}\,. \tag{8}$$

The formalism of Lagrangian Perturbation Theory allows us to find a solution to the divergence of equation (5), expanding the displacement vector $\mathbf{\Psi}$ as follows,

$$\mathbf{\Psi} = \mathbf{\Psi}^{(1)} + \mathbf{\Psi}^{(2)}\,. \tag{9}$$

We used the L-PICOLA code [20] to perform our simulations, this choice was made because L-PICOLA is considered a fast code, as it uses the Comoving Lagrangian Acceleration (COLA) methods, to solve the equation (5), which treads accuracy in small scales in order to gain computational speed, without sacrificing accuracy at large scales [21].

This code is based on a Particle Mesh scheme which calculates the gravitational force in the particles using a mesh, in general, those types of codes follow the next steps:

(i) Assign the particle charge to the mesh, by the Cloud-in-Cell linear interpolation, an explanation of this method can be found in Ref. [22].
(ii) Solve the Poisson equation (7) by means of the Fast Fourier Transform.
(iii) Calculate the gravitational force $\mathbf{F} = -\nabla\Phi$ in all the mesh points.
(iv) Using the Cloud-in-Cell interpolation get the force in each particle.
(v) Solve the equation of motion (5) in order to get the acceleration of the particles. The Kick-Drift/Leapfrog Method is used to update the particle velocities and positions. The process then starts again in step (i) above until the required evolution time is reached.

In comparison with other popular codes like GADGET [23, 24], that uses a hierarchical tree algorithm, and represents fluids by means of smoothed particle hydrodynamics, are necessary ~ 2000 timesteps in order to reach the same resolution that COLA gets in 10 timesteps.

1 The density contrast δ is also related to the MPS in the Fourier space through $P(k) \propto |\delta_k|^2$

The output snapshots of the cosmological simulations carry information about the final positions and velocities of the particles. In our case, we analyze them using the two-point correlation function, which can provide us with more information about the galaxies clustering in the Universe. The two-point correlation function $\xi(r)$ is defined as the probability excess, over random and averaged over all space, of finding two galaxies separated by a distance r (in volume elements δV_1 and δV_2) [25]

$$\delta P = n^2 \delta V_1 \delta V_2 [1 + \xi(r_{12})] \, . \tag{10}$$

We used the publicly-available CUTE code [26] to directly calculate the two-point correlation function from a given simulation snapshot.

We can also relate the two-point correlation function $\xi(r)$ with the mass power spectrum $P(k)$ by

$$\xi(r) = \frac{1}{2\pi^2} \int P(k) \frac{\sin kr}{kr} k^2 dk \, . \tag{11}$$

Eq. (11) shows the close relation between the two-point correlation function and the MPS, which allows us to use both of them to describe the clustering properties of the particles in a N-body simulation.

Firstly, we show in Fig. 2 the linear two-point correlation function $\xi(r)$ of the MPS presented before in Fig. 1; the correlation function was calculated using Eq. (11). Notice that the linear MPS does not take into account the non-linear features of structure formation that appear at late times, and then the obtained $\xi(r)$ should be considered just an approximation to the real thing. It can be seen that the different curves are indistinguishable from those of ΛCDM as long as the scalar field mass is larger than 10^{-25} eV. For the latter and smaller values, there appear some deviations at small scales that can be attributed to the cut-off scale in the MPS. For the latter, in the case $m = 10^{-25}$ eV, the power is suppressed for scales smaller than $r_c \sim 10\, h^{-1}\,\mathrm{Mpc}$ (which corresponds to the wavenumber $k_c \sim 0.1\, h\,\mathrm{Mpc}^{-1}$ such that $k_c r_c \sim 1$), and then the quantity $r^2 \xi$ should show an enhanced amplitude at around such scale. In contrast, the baryonic acoustic oscillations (BAO) scale is located in the range $r_{\mathrm{BAO}} \sim 100 - 120\, h^{-1}\,\mathrm{Mpc}$, for which $k_c r_{\mathrm{BAO}} \sim 10$, and then the result in the integral (11) is not affected by the cut-off in the MPS (unless we try with masses much lighter than 10^{-25} eV). Notice that the argument above is of general applicability, as it is based upon the mathematical properties of the correlation function (11), and then we expect the same results for other dark matter models that present a cut-off in their MPS, like in the case of warm dark matter [27–29].

Assuming the cosmological parameters given in the Ref. [1], we evolved the linear MPS shown in Fig. 1 in a box of size $[1024\,\mathrm{Mpc}/h]^3$, with $N = 256^3$ particles and $N_m = 768$ mesh points. Under these parameters, the size of the box is almost ten times larger than the BAO scale, and the number of mesh points is three times that of particles, which is recommended for L-PICOLA to provide optimum accuracy. In Fig. 3 a comparison between the linear and non-linear two-point correlation function is shown, for different scalar field values of boson mass and ΛCDM. From this comparison we can appreciate a discrepancy in the BAO scale for the linear and non-linear evolution, the BAO scale should not be affected by the non-linear evolution, therefore, it is necessary to do a BAO reconstruction [30] at that scale, in order to have a coincidence in both regimes.

4. Conclusions

N-body codes are one of the most helpful tools in cosmology to describe the process of structure formation. With the L-PICOLA code, we were able to get the non-linear evolution of a scalar field model for different boson masses, and the CUTE code allowed us to calculate the two-point correlation function from the simulation snapshots. The two-point correlation function shows

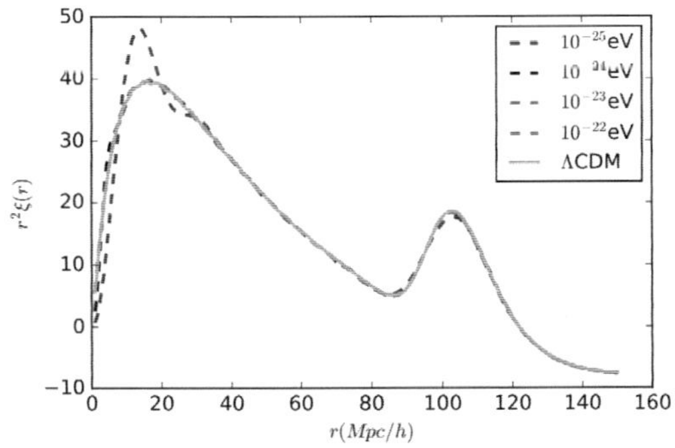

Figure 2. Linear two-point correlation function (that was also used as input in the L-PICOLA code) considering different values of the boson mass m. Notice that the different curves are indistiguishable from the result of standard cold dark matter model as long as $m > 10^{-25}$eV, but smaller values of m shows a noticeable discrepancy due to the existence of a sharp cut-off in the MPS, see also Fig. 1.

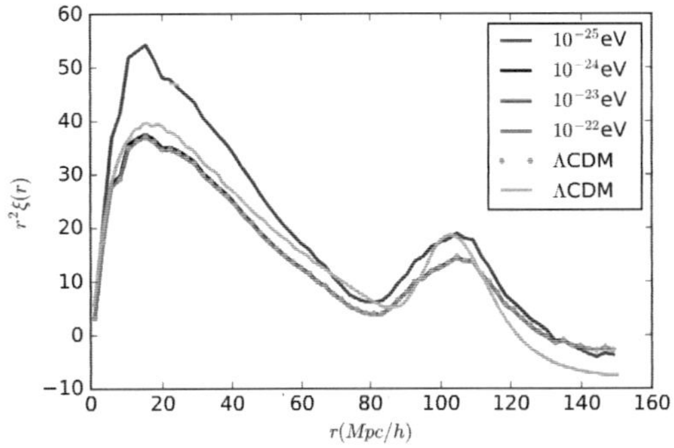

Figure 3. Comparison between the linear ΛCDM (yellow line, see also Fig. 2) and the nonlinear two-point correlation functions, for ΛCDM and different values of the boson mass. The nonlinear cases were obtained from a L-PICOLA simulation with particle number $N = 256^3$, mesh points $N_m = 768$, and a box size=$[1024\ \mathrm{Mpc}/h]^3$. It can be seen that both cases, linear and nonlinear, suggest that the boson mass should be $> 10^{-25}$ eV.

that the BAO scale is within the range $100 - 120\,\mathrm{Mpc}/h$ for all scalar field models. In addition, at large scales, a coincidence in the correlation function of the ΛCDM model and scalar field DM is expected, and this is actually the case for boson masses $m > 10^{-25}$eV in both regimes (linear and non-linear). This allows us to conclude that the scalar field model is a good alternative candidate for DM. However, at this point, we cannot get to distinguish between ΛCDM and scalar field at small scales, probably because of some error sources linked to the precision of the L-PICOLA code at those scales.

On the other hand, for boson masses smaller than 10^{-25} eV we observe an increase on the amplitude in the peaks of the two-point correlation function at small scales. Apart from the direct effect discussed above for Eq. (11), another reason may be that the cut-off in the linear MPS of the scalar field model is not properly interpreted by the codes L-PICOLA and CUTE, as they were originally designed for the ΛCDM model only. This can be remedied by appropriate changes in the available N-body codes [31], or by particular codes dedicated to scalar field models as those in Refs. [32–34]. Special attention should be paid to scales smaller than that of the BAO signal ($r < 140\,h^{-1}\,\mathrm{Mpc}$) where the differences with respect to CDM are more noticeable. The differences are easily calculated from the integral (11), but not quite so from the point of view of N-body simulations, as it is a difficult numerical task to obtain the large resolution required to model small galaxies in large enough cosmological volumes. This is left for future work that we shall present elsewhere.

Acknowledgments

We want to thank Alma X. González-Morales for help in the implementation of the numerical codes. RM-C acknowledges support from CONACyT México. This work was partially supported by PRODEP, DAIP-UGTO research grant 732/2017, CONACyT México under grants 167335, 179881, 269652 and Fronteras 281, Fundación Marcos Moshinsky, and the Instituto Avanzado de Cosmología (IAC) collaboration.

References

[1] Ade P A R *et al.* (Planck) 2015 (*Preprint* 1502.01589)
[2] Klypin A A, Kravtsov A V, Valenzuela O and Prada F 1999 *Astrophys. J.* **522** 82–92 (*Preprint* astro-ph/9901240)
[3] Du X, Behrens C and Niemeyer J C 2016 (*Preprint* 1608.02575)
[4] Moore B, Ghigna S, Governato F, Lake G, Quinn T R, Stadel J and Tozzi P 1999 *Astrophys. J.* **524** L19–L22 (*Preprint* astro-ph/9907411)
[5] 2016 (*Preprint* 1606.02835)
[6] de Blok W J G 2010 *Adv. Astron.* **2010** 789293 (*Preprint* 0910.3538)
[7] Boylan-Kolchin M, Bullock J S and Kaplinghat M 2011 *Mon. Not. Roy. Astron. Soc.* **415** L40 (*Preprint* 1103.0007)
[8] Boylan-Kolchin M, Bullock J S and Kaplinghat M 2012 *Mon. Not. Roy. Astron. Soc.* **422** 1203–1218 (*Preprint* 1111.2048)
[9] Papastergis E, Giovanelli R, Haynes M P and Shankar F 2015 *Astron. Astrophys.* **574** A113 (*Preprint* 1407.4665)
[10] Magana J and Matos T 2012 *J. Phys. Conf. Ser.* **378** 012012 (*Preprint* 1201.6107)
[11] Suárez A, Robles V H and Matos T 2014 *Astrophys. Space Sci. Proc.* **38** 107–142 (*Preprint* 1302.0903)
[12] Lee J W 2017 (*Preprint* 1704.05057)
[13] Hui L, Ostriker J P, Tremaine S and Witten E 2016 (*Preprint* 1610.08297)
[14] Matos T and Urena-Lopez L A 2001 *Phys. Rev.* **D63** 063506 (*Preprint* astro-ph/0006024)
[15] Matos T and Urena-Lopez L A 2000 *Class. Quant. Grav.* **17** L75–L81 (*Preprint* astro-ph/0004332)
[16] Ureña López L A and Gonzalez-Morales A X 2016 *JCAP* **1607** 048 (*Preprint* 1511.08195)
[17] Linares Cedeño F X, González-Morales A X and Arturo Ureña López L 2017 *Phys. Rev.* **D96** 061301 (*Preprint* 1703.10180)
[18] Ma C P and Bertschinger E 1995 *Astrophys. J.* **455** 7–25 (*Preprint* astro-ph/9506072)
[19] Bernardeau F, Colombi S, Gaztanaga E and Scoccimarro R 2002 *Phys. Rept.* **367** 1–248 (*Preprint* astro-ph/0112551)
[20] Howlett C, Manera M and Percival W J 2015 (*Preprint* 1506.03737)
[21] Tassev S, Zaldarriaga M and Eisenstein D 2013 *JCAP* **1306** 036 (*Preprint* 1301.0322)
[22] Birdsall C K and Fuss D 1997 *Journal of Computational Physics* **135** 141–148
[23] Springel V, Yoshida N and White S D M 2001 *New Astron.* **6** 79 (*Preprint* astro-ph/0003162)
[24] Springel V 2005 *Mon. Not. Roy. Astron. Soc.* **364** 1105–1134 (*Preprint* astro-ph/0505010)
[25] Kolb E W and Turner M S 1990 *Front. Phys.* **69** 1–547
[26] Alonso D 2012 (*Preprint* 1210.1833) URL http://members.ift.uam-csic.es/dmonge/CUTE
[27] Colombi S, Dodelson S and Widrow L M 1996 *Astrophys. J.* **458** 1 (*Preprint* astro-ph/9505029)
[28] Viel M, Markovic K, Baldi M and Weller J 2012 *Mon. Not. Roy. Astron. Soc.* **421** 50–62 (*Preprint* 1107.4094)
[29] Smith R E and Markovic K 2011 *Phys. Rev.* **D84** 063507 (*Preprint* 1103.2134)
[30] Padmanabhan N, Xu X, Eisenstein D J, Scalzo R, Cuesta A J, Mehta K T and Kazin E 2012 *Mon. Not. Roy. Astron. Soc.* **427** 2132–2145 (*Preprint* 1202.0090)
[31] Corasaniti P S, Agarwal S, Marsh D J E and Das S 2017 *Phys. Rev.* **D95** 083512 (*Preprint* 1611.05892)
[32] Schive H Y, Chiueh T and Broadhurst T 2014 *Nature Phys.* **10** 496–499 (*Preprint* 1406.6586)
[33] Schive H Y, Liao M H, Woo T P, Wong S K, Chiueh T, Broadhurst T and Hwang W Y P 2014 *Phys. Rev. Lett.* **113** 261302 (*Preprint* 1407.7762)
[34] Zhang J, Tsai Y L S, Cheung K and Chu M C 2016 (*Preprint* 1611.00892)

IOP Conf. Series: Journal of Physics: Conf. Series **1030** (2018) 012008 doi:10.1088/1742-6596/1030/1/012008

Series solutions of single-field models of inflation

V M Jaramillo-Pérez[1], I Padilla-Gay[1], A Diez-Tejedor[2], L A Ureña-López[2]

[1] División de Ciencias e Ingenierías, Campus León, Universidad de Guanajuato, 37150, León, Guanajuato, México

[2] Departamento de Física, DCI, Campus León, Universidad de Guanajuato, 37150, León, Guanajuato, México

E-mail: `lurena@ugto.mx`

Abstract. We describe a generic method, first presented in Ref.[1], to find inflationary solutions without the need to be specific about the scalar potential. The method shows that single-field models of inflation can be classified in two groups according to their predictions of inflationary observables, we call these groups as Class I and Class II. We apply the method to the study of different scalar potentials and compare their results with recent observational data. We conclude that potentials belonging to Class I are in better agreement with observations mostly because they can provide large enough values of the spectral index of linear perturbations.

1. Introduction

What we actually call the standard model of Cosmology is a phenomenological description of the physics that seems to be at play to reproduce the observed Universe [2, 3, 4, 5]. The predictions of such model are in strikingly good agreement with a variety of observational constraints [6], and it is because of this reason that it is the fiducial model used in present and future observational forecasts. The standard cosmological model is also known as the ΛCDM model, for the initials of its main matter components at the present time: the cosmological constant Λ and Cold Dark Matter (CDM). The core of the model is based on a spatially-flat expanding Universe whose gravitational dynamics is governed by the equations of motion of General Relativity (for a pedagogical introduction to the state of the art of modern cosmology please see Ref. [7]).

Apart form this, the ΛCDM model also needs to set up the appropriate initial conditions so that the formation of structure can start as required by present observations of the distribution of galaxies. These seeds of structure formation must have been produced by some separate process in the early Universe. It is because of this that a period of inflation, which refers to an accelerated stage in the very early Universe, remains as one of the cornerstone ideas of modern Cosmology [8, 9, 10].

In the case of inflationary models with a single scalar field, the equations of motion can be simplified because of the assumption that the scalar field is the only matter field present in the early Universe. Another simplification is that we do not need to solve the equations of motion for linear perturbations, as there already exist general results for the features of the primordial perturbations that only require input from the dynamics in the background [8, 9, 10, 11].

In this paper we extend the study first presented in Ref. [1] to consider more examples of inflationary potentials, in particular some selected examples from the so-called Enciclopaedia

Inflationaris [12]. We focus our attention in those models that are able to provide an initial de Sitter phase of inflation, and calculate their inflationary predictions to compare them with the general solutions presented in Ref. [12].

A brief description of the paper is as follows. In Sec. 2 we describe the general mathematical framework to describe inflationary models and the series technique to find their solutions. We also present the results obtained for the inflationary numbers in the case of selected models and make a classification of them according to the classes suggested in Ref. [1]. Finally, Sec. 3 is devoted to conclusions.

2. Mathematical background

We will be concerned here with the inflationary dynamics of a (homogeneous) scalar field ϕ endowed with a scalar field potential $V(\phi)$. If we take the Friedmann-Robertson-Walker line element for a homogeneous and isotropic flat universe $ds^2 = -dt^2 + a^2(t)\delta_{ij}dx^i dx^j$, where $a(t)$ is the scale factor, the equations of motion we have to solve are

$$H^2 = \frac{\kappa^2}{3}\left[\frac{1}{2}\dot{\phi}^2 + V(\phi)\right], \quad \dot{H} = -\kappa^2\dot{\phi}^2, \quad \ddot{\phi} = -3H\dot{\phi} - \partial_\phi V. \tag{1}$$

Next, we define the following transformation of variables,

$$\frac{\kappa\dot{\phi}}{\sqrt{6}H} = \sin(\theta/2), \quad \frac{\kappa V^{1/2}}{\sqrt{3}H} = \cos(\theta/2), \quad y_1 = -2\sqrt{2}\frac{\partial_\phi V^{1/2}}{H}, \quad y_2 = -\frac{4\sqrt{3}}{\kappa}\frac{\partial_\phi^2 V^{1/2}}{H}, \tag{2}$$

under which the dynamical equations in (1) can be written generically as

$$\theta' = -3\sin\theta + y_1, \quad y_1' = \frac{3}{2}(1-\cos\theta)\, y_1 + \sin(\theta/2) y_2. \tag{3}$$

Here, a prime denotes derivative with respect to the number of e-folds of expansion $N \equiv \ln(a)$, and the Friedmann constraint appears as a consequence of the definitions in Eq. (2).

The system of equations (3) is a neat representation of the original equations of motion (1) that only requires input information about the second potential variable y_2. It must be noticed that the new angular variable θ has a direct interpretation in terms of the scalar field equation of state (EoS) w_ϕ, which is defined as

$$w_\phi = \frac{\frac{1}{2}\dot{\phi}^2 - V(\phi)}{\frac{1}{2}\dot{\phi}^2 + V(\phi)} = -\cos\theta. \tag{4}$$

In terms of the EoS, the inflationary phase is represented by an evolution form $w_\phi \simeq -1$ to $w_\phi = -1/3$. The latter can be written in terms of the angular variable as $\theta : 0 \to \theta_{end}$, where $\theta_{end} = \arccos(1/3) = 2\arcsin(1/\sqrt{3}) \simeq 1.2309\ldots$.

Following the work in Ref. [1], we will try a series solution of the second expression in Eq. (3) with the following ansatz,

$$y_1 = \sum_{j=1} k_{1j}\theta^j, \tag{5}$$

where k_{1j} are all constant coefficients. If the values of k_{1j} can be calculated somehow, then a solution of Eqs. (3) can be found from the series expansion

$$\theta' = (k_{11} - 3)\theta + k_{12}\theta^2 + (k_{13} + 1/2)\theta^3 + \ldots. \tag{6}$$

Eq. (6) can then be solved at any order to provide a solution in the form $\theta_N = \theta_N(k_{1j}, \theta_{end}, N)$, where the subscript N denotes the number of e-foldings before the end of inflation. Actually, the series solution up to the third order θ^3 is

$$(k_{11} - 3)N^{(3rd)} = \ln\left(\frac{\theta_{end}}{\theta_N}\right) + \frac{\theta_-}{\theta_+ - \theta_-}\ln\left(\frac{\theta_{end} - \theta_+}{\theta_N - \theta_+}\right) - \frac{\theta_+}{\theta_+ - \theta_-}\ln\left(\frac{\theta_{end} - \theta_-}{\theta_N - \theta_-}\right), \quad (7)$$

where

$$\theta_\pm = \frac{-k_{12} \pm \sqrt{k_{12}^2 - 2(k_{11} - 3)(2k_{13} + 1)}}{2k_{13} + 1}, \quad \theta_1 = \frac{k_{11} - 3}{k_{12}}. \quad (8)$$

Eq. (7) is the expression that we will use to calculate the solutions and inflationary quantities for any given model. Approximate formulas can be obtained from Eq. (7) for special cases in which either or both of the following conditions $k_{11} = 3$ and $k_{12} = 0$ apply.

As shown in Ref. [1], the coefficients k_{1j} can be systematically determined if the scalar field potential allows a phase of the so-called de Sitter inflation, which is characterized by $w_\phi \simeq -1$; this is a regime that exists as long as the scalar field potential is the dominant term of the energy density at very early times. The coefficients k_{1j} are then calculated from appropriate combinations of the field derivatives of $V(\phi)$ at the point of de Sitter inflation. In this form Eq. (7) gives the solution that corresponds to the given scalar field potential.

For the calculation of inflationary quantities, we choose the so-called Hubble slow-roll (HSR) variables [13], under which various inflationary observables can be written in terms of the so-called HSR parameters ϵ_H and η_H. For instance, the spectral index n_s and the tensor-to-scalar ratio r are given at first order in the HSR parameters as:

$$1 - n_s = 4\epsilon_H - 2\eta_H = 12\sin^2(\theta/2) - 6 + y_1\cot(\theta/2), \quad (9)$$

$$r = 16\epsilon_H = 48\sin^2(\theta/2), \quad (10)$$

where the last expressions on the right hand side appear from the change to polar variables (2). Taking into account the expansions in Eq. (5), the expressions in Eqs. (9) and (10) can be written alternatively as:

$$1 - n_s \simeq 2(k_{11} - 3) + 2k_{12}\theta_N + (2k_{13} - k_{11}/6 + 3)\theta_N^2, \quad r \simeq 12\theta_N^2. \quad (11)$$

In principle, Eqs. (7) and (11) is all what we need to find the values of the inflationary observables n_s and r, once the values of the expansion coefficients k_{1j} have been determined by other means. Selected examples from the Enciclopaedia Inflationaris [12] are shown in Table 1, which extend the number of cases originally shown in Ref. [1], together with their corresponding values of the series coefficients k_{1j} and the obtained values of the inflationary quantities. The models are also classified as Class I ($k_{12} = 0$) and Class II ($k_{12} \neq 0$), which are the general classes suggested in Ref. [1].

To finish this section, we show in Fig. 1 the generic predictions for the observables n_s and r of the two classes of inflationary solutions, for $N = 50$, which is a typical number of e-folds before the end of inflation. When compared with the updated observational constraints reported in Ref. [14], see also Ref. [15]: $n_s = 0.9783 \pm 0.0072$ and $r < 0.074$, it can be seen that it is Class I, rather than Class II, the one set of models that can provide a good fit to the observational data. This is generically explained due to the fact that, for $N = 50$, $n_s \leq 0.98$ ($n_s \leq 0.96$) for models in Class I (Class II). Even more, for the models in Class I a better agreement with observations is reached when $k_{11} \to 3$ and $k_{13} \to \infty$, as in this case the prediction for the spectral index is just $n_s = 1 - 1/N$, and then $n_s = 0.98$ if $N = 50$ (see Ref. [1] for more details).

Table 1. List of selected models of inflation taken from the Enciclopaedia Inflationaris [12]. The columns indicate, in order from left to right: 1) Name of the potential, 2) Functional form of the potential, 3) de Sitter point; 4-6) Series coefficients k_{1j}, see Eq. (5); 7) Class of the model; 8) Spectral index n_s; and 9) Tensor-to-scalar ratio r, see Eq. (11).

Model	Potential	$\phi_{\rm dS}$	k_{11}	k_{12}	k_{13}	Class	n_s	r
RCHI	$1-2e^{\sqrt{2/3}\phi\kappa}-\frac{\kappa A_I\phi}{16\pi^2\sqrt{6}}$	$-\infty$	3	0	1	I	0.970	8.0×10^{-2}
MLFI	$\kappa^2\phi^2\left(1+\alpha\kappa^2\phi^2\right)$	$-\infty$	3	0	-1/8	I	0.941	3.1×10^{-1}
HI	$\left(1-e^{\sqrt{2/3}\kappa\phi}\right)^2$	$-\infty$	3	1	$-1/12$	II	0.962	4.6×10^{-3}
ESI	$1-e^{0.1\kappa\phi}$	$-\infty$	3	0.122	0.995	II	1.058	2.4×10^{-1}

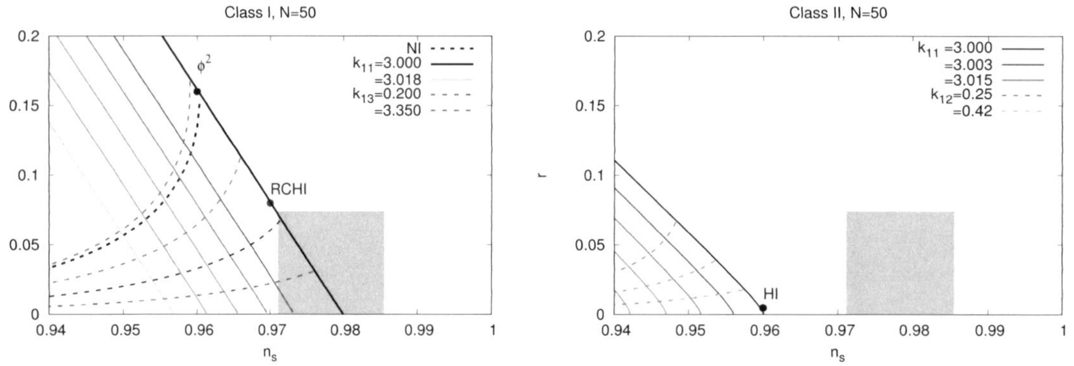

Figure 1. General results on the plane n_s vs r for the Class I (left panel) and II (right panel) of the single field models of inflation, for different values of the coefficients k_{11}, k_{12} and k_{13}. The plots show the results for $N = 50$. Notice that Class I is not completely ruled out by observations, mostly because it can provide low enough values of the tensor-to-scalar ratio r. The predictions from the typical Large Field Inflation for the potential ϕ^2 [16], Natural Inflation (NI) [17] and Radiatively Corrected Higgs Inflation (RCHI)[12], are indicated on the left panel for comparison, whereas we do the same for the Higgs Inflation (HI) model [18] on the right panel. The predictions from other potentials listed in Table 1 lie outside of the regions shown in the figures, and are then also incompatible with the observational constraints. The grey shaded rectangle in each figure represents the observational constraints as reported recently in Ref. [14].

3. Conclusions

We have presented an extended study of single-field models of inflation following the prescription in Ref. [1]. In doing so, we have shown that the series solution and the general classification of models can be applied to models beyond those presented in the previous reference.

However, we have also discovered that the series method may require the consideration of higher order solutions, mostly because the influence of some free parameters in the inflationary models is not present, in some instances, in the lowest order coefficients. This should be done specially for some models in Class I (with $k_{11} = 3$ and $k_{12} = 0$), for which one could even find an analytical solution up to the fifth order in Eq. (6). It is then expected that the higher-order solutions will provide more accurate predictions for the inflationary quantities. This is work in progress that we shall present elsewhere.

IOP Conf. Series: Journal of Physics: Conf. Series **1030** (2018) 012008 doi:10.1088/1742-6596/1030/1/012008

Acknowledgments
VMJ-P and IP-G acknowledge support from CONACyT México and PRODEP. This work was partially supported by PRODEP, DAIP-UGTO research grant 732/2017 and 878/2017, PIFI, CONACyT México under grants 167335, 179881, 269652 and Fronteras 281, Fundación Marcos Moshinsky, and the Instituto Avanzado de Cosmología (IAC) collaboration.

References
[1] Urena-Lopez L A 2016 *Phys. Rev.* **D94** 063532 (*Preprint* 1512.07142)
[2] Kolb E W and Turner M S 1990 *The Early Universe* (Frontiers in Physics)
[3] Peebles P J E 1993 *Principles of Physical Cosmology* (Princeton Series in Physics)
[4] Dodelson S 2003 *Modern Cosmology* (Academic Press)
[5] Weinberg S 2008 *Cosmology* (Oxford University Press)
[6] Ade P A R *et al.* (Planck) 2016 *Astron. Astrophys.* **594** A13 (*Preprint* 1502.01589)
[7] Liddle A R 2015 *Modern Cosmology* (Wiley-Blackwell)
[8] Liddle A R and Lyth D H 2000 *Cosmological inflation and large scale structure* (Cambridge University Press)
[9] Mukhanov V 2005 *Physical Foundations of Cosmology* (Cambridge University Press)
[10] Bassett B A, Tsujikawa S and Wands D 2006 *Rev. Mod. Phys.* **78** 537–589 (*Preprint* astro-ph/0507632)
[11] Vennin V, Koyama K and Wands D 2015 *JCAP* **1511** 008 (*Preprint* 1507.07575)
[12] Martin J, Ringeval C and Vennin V 2014 *Phys.Dark Univ.* (*Preprint* 1303.3787)
[13] Liddle A R, Parsons P and Barrow J D 1994 *Phys.Rev.* **D50** 7222–7232 (*Preprint* astro-ph/9408015)
[14] Guo R Y and Zhang X 2017 (*Preprint* 1704.04784)
[15] Ade P A R *et al.* (Planck) 2016 *Astron. Astrophys.* **594** A20 (*Preprint* 1502.02114)
[16] Linde A D 1983 *Phys.Lett.* **B129** 177–181
[17] Freese K and Kinney W H 2015 *JCAP* **1503** 044 (*Preprint* 1403.5277)
[18] Starobinsky A A 1980 *Phys.Lett.* **B91** 99–102

IOP Conf. Series: Journal of Physics: Conf. Series **1030** (2018) 012009 doi:10.1088/1742-6596/1030/1/012009

A simple model of universe with a polytropic equation of state

Pierre-Henri Chavanis

Laboratoire de Physique Théorique, Université Paul Sabatier, 118 route de Narbonne, 31062 Toulouse, France

E-mail: `chavanis@irsamc.ups-tlse.fr`

Abstract. We construct a simple model of universe with a generalized equation of state $p = (\alpha + k\rho^{1/n})\rho c^2$ having a linear component $p = \alpha\rho c^2$ and a polytropic component $p = k\rho^{1+1/n}c^2$. For $\alpha = 1/3$, $n = 1$ and $k = -4/(3\rho_P)$, where $\rho_P = 5.16 \times 10^{99}\,\mathrm{g/m^3}$ is the Planck density, the equation of state $p = \frac{1}{3}(1 - 4\rho/\rho_P)\rho c^2$ provides a model of early universe without singularity describing the transition between the vacuum energy era (Planck era) and the radiation era. The universe starts from $t = -\infty$ but, when $t < 0$, its size is less than the Planck length $l_P = 1.62 \times 10^{-35}\,\mathrm{m}$, requiring a quantum gravity treatment. For $t \geq 0$, the universe undergoes an accelerated expansion (early inflation) that brings it from the Planck size $l_P = 1.62 \times 10^{-35}\,\mathrm{m}$ to a size $a_1 = 2.61 \times 10^{-6}\,\mathrm{m}$ on a timescale of about 23.3 Planck times $t_P = 5.39 \times 10^{-44}\,\mathrm{s}$. When $t > t_1 = 23.3 t_P = 1.25 \times 10^{-42}\,\mathrm{s}$, the universe decelerates and enters in the radiation era. For $\alpha = 0$, $n = -1$ and $k = -\rho_\Lambda$, where $\rho_\Lambda = 7.02 \times 10^{-24}\,\mathrm{g/m^3}$ is the cosmological density, the equation of state $p = -\rho_\Lambda c^2$ describes the transition from a decelerating universe dominated by dark matter to an accelerating universe (late inflation) dominated by dark energy (de Sitter era). This transition takes place at a size $a_2 = 0.204 l_\Lambda = 8.95 \times 10^{25}\mathrm{m}$ corresponding to a time $t_2 = 0.203 t_\Lambda = 2.97 \times 10^{17}\mathrm{s}$ where $l_\Lambda = 4.38 \times 10^{26}\mathrm{m}$ is the cosmological length and $t_\Lambda = 1.46 \times 10^{18}\,\mathrm{s}$ the cosmological time. The present universe turns out to be just at the transition ($t_0 \sim t_2$). Our model generalizes the standard ΛCDM model by incorporating naturally a phase of early inflation that avoids the primordial singularity. Furthermore, it reveals a nice "symmetry" between the early universe (vacuum energy + radiation) and the late universe (dark matter + dark energy). They are described by two polytropic equations of state with index $n = +1$ and $n = -1$ respectively. Furthermore, the cosmological constant Λ in the late universe plays a role similar to the Planck constant $\hbar$ in the early universe. The mathematical formulae in the early and in the late universe are strikingly symmetric. We interpret the cosmological constant as a fundamental constant of nature describing the "cosmophysics" just like the Planck constant describes the "microphysics". The Planck density and the cosmological density represent fundamental upper and lower bounds differing by 123 orders of magnitude. They are responsible for a phase of inflation in the early and late universe. The cosmological constant "problem" may be a false problem. Our model does not present any singularity and exists eternally in the past and in the future (aioniotic universe). On the other hand, it admits a scalar field interpretation based on an inflaton, quintessence, or tachyonic field. This correspondence is interesting because the early inflation and the late acceleration of the universe are usually described in terms of a scalar field. We determine the potential and the mass of this scalar field both in the early and late universe.

1. Introduction

According to contemporary cosmology, the present energy content of the universe is composed of approximately 5% baryonic matter, 20% dark matter, and 75% dark energy. The expansion of the universe began in a tremendous inflationary burst driven by vacuum energy (Planck era). Between 10^{-35} and 10^{-33} seconds after the Big Bang, the universe expanded by a factor 10^{30} [1]. The universe then entered in the radiation era and, when the temperature cooled down below approximately 10^3 K, in the matter era [2]. At present, it undergoes an accelerated expansion (de Sitter era) presumably due to the cosmological constant Λ or to some form of dark energy [3]. This corresponds to a second period of inflation. Despite the success of the standard cold dark matter model (ΛCDM), the nature of vacuum energy, dark matter and dark energy, remains very mysterious and leads to many speculations.

The phase of inflation in the early universe, which is necessary to solve notorious difficulties such as the singularity problem, the flatness problem (or the fine-tuning problem) and the horizon problem (or the causality problem) and to account for the observed spectrum of primordial density fluctuations, is usually described by some hypothetical scalar field ϕ with its origin in the quantum fluctuations of vacuum [1]. This leads to an equation of state $p = -\rho c^2$ with a negative pressure, implying a constant energy density, called the vacuum energy. This energy density is usually identified with the Planck density $\rho_P = c^5/G^2\hbar = 5.16 \times 10^{99}\,\mathrm{g/m^3}$. As a result of the vacuum energy, the universe expands exponentially rapidly on a timescale of the order of the Planck time $t_P = 1/(G\rho_P)^{1/2} = (\hbar G/c^5)^{1/2} = 5.39 \times 10^{-44}$s. This phase of early inflation (Planck era) is followed by the radiation era described by an equation of state $p = \rho c^2/3$ ($\rho \propto a^{-4}$, $a \propto t^{1/2}$ and $\rho = 3/32\pi G t^2$) and by the matter era described by an equation of state $p = 0$ ($\rho \propto a^{-3}$, $a \propto t^{2/3}$ and $\rho = 1/6\pi G t^2$).

The phase of acceleration in the late universe, which has been revealed by the observations of distant type Ia supernovae is usually ascribed to the cosmological constant Λ which is equivalent to a constant energy density $\rho_\Lambda = \Lambda/(8\pi G) = 7.02 \times 10^{-24}\,\mathrm{g/m^3}$ called the dark energy [3]. This acceleration can be modeled by an equation of state $p = -\rho c^2$ with a negative pressure, implying a constant energy density identified with the cosmological density ρ_Λ. As a result of the dark energy, the universe expands exponentially rapidly on a timescale of the order of the cosmological time $t_\Lambda = 1/(G\rho_\Lambda)^{1/2} = (8\pi/\Lambda)^{1/2} = 1.46 \times 10^{18}$ s. This leads to a phase of late inflation (de Sitter era). Instead of introducing a cosmological constant, it has been proposed to explain the acceleration of the universe in terms of a dark energy with a time-varying density. Inspired by the analogy with the early inflation, some authors have represented the dark energy by a scalar field called quintessence [4]. As an alternative to quintessence, other authors have proposed to model the acceleration of the universe by an exotic fluid with an equation of state of the form $p/c^2 = -A/\rho$ called the Chaplygin gas [5]. At late times, this equation of state leads to a constant energy density implying an exponential growth of the scale factor that is similar to the effect of the cosmological constant. At earlier times, this equation of state returns the results of the CDM model ($p \simeq 0$) where the density decreases as $\rho \propto a^{-3}$ (Einstein-de Sitter solution). Therefore, it provides a unification of dark matter ($\rho \propto a^{-3}$) and dark energy ($\rho = \rho_\Lambda$) in the late universe. Furthermore, it gives a real speed of sound which is non-trivial for fluids with negative pressure. The Chaplygin gas has some connection with string theory and can be obtained from the Nambu-Goto action for d-branes moving in a $(d+2)$-dimensional spacetime in the light-cone parametrization. However, the Chaplygin gas model does not give a good description of the evolution of the universe between the matter era and the dark energy era [6, 7].

In this paper, we construct a simple model of universe taking into account the effect of vacuum energy, radiation, dark matter and dark energy through a generalized equation of state of the form

$$p = (\alpha\rho + k\rho^{1+1/n})c^2. \tag{1}$$

This is the sum of a standard linear equation of state $p = \alpha\rho c^2$ and a polytropic equation of state $p = k\rho^\gamma c^2$, where k is the polytropic constant and $\gamma = 1 + 1/n$ the polytropic index. An exhaustive study of this equation of state, considering all the possible cases for arbitrary values of α, k and n, is given in [8, 9]. We have found the following structure. Positive indices $n > 0$ describe the early universe where the polytropic component dominates the linear component because the density is high. Negative indices $n < 0$ describe the late universe where the polytropic component dominates the linear component because the density is low. On the other hand, a positive polytropic pressure ($k > 0$) leads to past or future singularities (or peculiarities) while a negative polytropic pressure ($k < 0$) leads to a phase of exponential expansion (inflation) in the past or in the future. The case of a phantom universe where the density increases with the scale factor has been studied in [10].

In this paper, we show that the early and late evolution of the universe can be described "symmetrically" by two polytropic equations of state of the form of Eq. (1) with index $n = 1$ and $n = -1$, respectively. Our model [see Eq. (57)] generalizes the standard ΛCDM model by removing the primordial singularity and accounting for a phase of early inflation. We study the dynamical and thermodynamical evolution of the universe and determine a generalized Stefan-Boltzmann law [see Eq. (13)] valid in the very early universe, before the radiation era. On the other hand, we show that our model admits a scalar field interpretation based on an inflaton, quintessence, or tachyonic field. This correspondence is interesting because the early inflation and the late acceleration of the universe are usually described in terms of a scalar field. We determine the potential and the mass of this scalar field both in the early [see Eqs. (19) and (25)] and late [see Eqs. (38), (42) and (49)] universe.

2. Basic equations of cosmology

We assume that the universe is isotropic and homogeneous at large scales and contains a uniform perfect fluid of energy density $\epsilon(t) = \rho(t)c^2$ and pressure $p(t)$. We also assume that the universe is flat in agreement with observations of the cosmic microwave background (CMB). It that case, the Einstein equations reduce to

$$\frac{d\rho}{dt} + 3\frac{\dot{a}}{a}\left(\rho + \frac{p}{c^2}\right) = 0, \tag{2}$$

$$H^2 = \left(\frac{\dot{a}}{a}\right)^2 = \frac{8\pi G}{3}\rho, \tag{3}$$

where $a(t)$ is the scale factor (by an abuse of language, we shall sometimes call it the "radius" of the universe) and $H = \dot{a}/a$ is the Hubble parameter. These are the well-known Friedmann equations describing an expanding universe [2]. The first equation is the energy equation (it can also be viewed as an equation of continuity). For a given barotropic equation of state $p = p(\rho)$, it determines the relation between the energy density and the scale factor. Then, the temporal evolution of the scale factor is given by Eq. (3). We have not written the cosmological constant Λ because its effect will be taken into account in the equation of state.

The equation of state parameter w, the deceleration parameter q and the speed of sound c_s are defined by

$$w = \frac{p}{\rho c^2}, \qquad q(t) = -\frac{\ddot{a}a}{\dot{a}^2} = \frac{1 + 3w(t)}{2}, \qquad c_s^2 = p'(\rho), \tag{4}$$

where the second equality in the expression of q, obtained from the Friedmann equations, is valid for a flat universe. The universe is decelerating when $q > 0$ (or $w > -1/3$) (strong energy condition) and accelerating when $q < 0$ (or $w < -1/3$). On the other hand, according to Eq. (2), the density decreases with the scale factor if $w > -1$ (null dominant energy condition) and increases with the scale factor if $w < -1$. The latter case corresponds to a phantom universe [11].

We will also need the thermodynamic equation

$$\frac{dp}{dT} = \frac{1}{T}(\rho c^2 + p), \tag{5}$$

which can be derived from the first principle of thermodynamics [2]. For a given barotropic equation of state $p = p(\rho)$, this equation can be integrated to obtain the relation $T = T(\rho)$ between the temperature and the density. Combining Eqs. (2) and (5) we get

$$\frac{dS}{dt} = 0, \qquad \text{where} \qquad S = \frac{a^3}{T}(p + \rho c^2) \tag{6}$$

is the entropy of the universe in a volume a^3. This shows that the Friedmann equations imply the conservation of the entropy.

To close the system of equations, we have to specify an equation of state. We shall consider a generalized equation of state of the form of Eq. (1) in the early and late universe [8, 9, 12].

3. Early universe: vacuum energy + radiation

3.1. Equation of state

We first consider the equation of state (1) with $\alpha = 1/3$, $n = 1$ and $k = -4/(3\rho_P)$ which provides a model of early universe dominated by vacuum energy and radiation. This equation of state can be rewritten as [8, 12]:

$$p = \frac{1}{3}\left(1 - \frac{4\rho}{\rho_P}\right)\rho c^2. \tag{7}$$

The continuity equation (2) can be integrated into

$$\rho = \frac{\rho_P}{(a/a_1)^4 + 1}, \tag{8}$$

where a_1 is a constant of integration.

The equation of state parameter, the deceleration parameter, and the speed of sound are given by

$$w = \frac{1}{3} - \frac{4}{3}\frac{\rho}{\rho_P}, \qquad q = 1 - 2\frac{\rho}{\rho_P}, \qquad \frac{c_s^2}{c^2} = \frac{1}{3} - \frac{8}{3}\frac{\rho}{\rho_P}. \tag{9}$$

As the universe expands from $a = 0$ to $a = +\infty$, the density decreases from ρ_P to 0, the equation of state parameter w increases from -1 to $1/3$, the deceleration parameter q increases from -1 to 1, and the ratio $(c_s/c)^2$ increases from $-7/3$ to $1/3$.

3.2. Vacuum energy era: early inflation

When $a \ll a_1$, the density is approximately constant ($\rho \simeq \rho_P$) and the pressure tends to the value $p = -\rho_P c^2$. Since this solution is expected to describe the very early universe, it is natural to identify the constant ρ_P with the Planck density $\rho_P = c^5/G^2\hbar = 5.16 \times 10^{99}\,\mathrm{g/m^3}$ (vacuum energy) which may represent a fundamental upper bound for the density. This is how quantum mechanics is taken into account in our model. A constant value of the density gives rise to a phase of early inflation. From the Friedmann equation (3), we find that the scale factor increases exponentially rapidly with time as

$$a(t) \sim l_P e^{(8\pi/3)^{1/2} t/t_P}, \tag{10}$$

where $t_P = 1/(G\rho_P)^{1/2} = (\hbar G/c^5)^{1/2} = 5.39 \times 10^{-44}$s is the Planck time which gives the timescale of the exponential growth. We have defined the time $t = 0$ such that $a(0)$ is equal to the Planck length $l_P = ct_P = (G\hbar/c^3)^{1/2} = 1.62 \times 10^{-35}$ m. Mathematically speaking, this universe exists at any time in the past ($a \to 0$ and $\rho \to \rho_P$ for $t \to -\infty$), so there is no primordial singularity. For $t < 0$, the radius of the universe is smaller than the Planck length. However, when $a \to 0$, we cannot ignore the quantum fluctuations associated with the spacetime. In that case, we cannot use the classical Einstein equations anymore and a theory of quantum gravity is required. It is not known whether quantum gravity will remove, or not, the primordial singularity. Therefore, we cannot extrapolate the solution (10) to the infinite past. However, this solution may provide a *semi-classical* description of the phase of early inflation when $a \geq l_P$.

3.3. Radiation era

When $a \gg a_1$, we get $\rho/\rho_P \sim (a_1/a)^4$. When the density is "low", the equation of state (7) reduces to the linear equation of state $p = \rho c^2/3$ which describes the radiation era. The conservation of $\rho_{\rm rad}\, a^4$ implies that $\rho_P a_1^4 = \rho_{\rm rad,0}\, a_0^4$ where $\rho_{\rm rad,0}$ is the present density of radiation and $a_0 = c/H_0 = 1.32 \times 10^{26}$ m the present distance of cosmological horizon determined by the Hubble constant $H_0 = 2.27 \times 10^{-18}\,{\rm s}^{-1}$ (the Hubble time is $H_0^{-1} = 4.41 \times 10^{17}$s). Writing $\rho_{\rm rad,0} = \Omega_{\rm rad,0}\rho_0$ where $\rho_0 = 3H_0^2/8\pi G = 9.20 \times 10^{-24}\,{\rm g/m^3}$ is the present density of the universe (the present mass of the universe is $M_0 = \rho_0 a_0^3 = 3c^3/8\pi G H_0 = 2.11 \times 10^{55}$ g) and $\Omega_{\rm rad,0} = 8.48 \times 10^{-5}$ is the present fraction of radiation in the universe, we obtain $a_1/a_0 = (\Omega_{\rm rad,0}\rho_0/\rho_P)^{1/4}$, hence $a_1 = 2.61 \times 10^{-6}$ m. This scale is intermediate between the Planck length l_P and the present size of the universe a_0 ($a_1/l_P = 1.61 \times 10^{29}$ and $a_1/a_0 = 1.97 \times 10^{-32}$). It marks the transition between the vacuum energy era and the radiation era. When $a \gg a_1$, the Friedmann equation (3) yields $a/a_1 \sim (32\pi/3)^{1/4}(t/t_P)^{1/2}$ and $\rho/\rho_P \sim (3/32\pi)(t_P/t)^2$.

3.4. The general solution

The general solution of the Friedmann equation (3) with Eq. (8) is [8, 12]:

$$\sqrt{(a/a_1)^4+1} - \ln\left(\frac{1+\sqrt{(a/a_1)^4+1}}{(a/a_1)^2}\right) = 2\left(\frac{8\pi}{3}\right)^{1/2}\frac{t}{t_P} + C, \tag{11}$$

where $C \simeq -134$ is a constant of integration determined such that $a = l_P$ at $t = 0$. For $t \to -\infty$, Eq. (11) returns Eq. (10) with an excellent approximation. The transition between the vacuum energy era and the radiation era ($a = a_1$) corresponds to a density $\rho_1 = \rho_P/2$ and a time $t_1 = (3/32\pi)^{1/2}[\sqrt{2} - \ln(1+\sqrt{2}) - C]t_P$. The universe is accelerating when $a < a_c$ (i.e. $\rho > \rho_c$) and decelerating when $a > a_c$ (i.e. $\rho < \rho_c$) where $\rho_c = \rho_P/2$ and $a_c = a_1$. The time t_c at which the universe starts decelerating corresponds to the time at which the curve $a(t)$ presents a first inflexion point. It turns out that this inflexion point coincides with a_1 so it also marks the end of the inflation ($t_c = t_1$).

3.5. The pressure

The pressure is given by Eq. (7). Using Eq. (8), we get

$$p = \frac{\frac{1}{3}(a/a_1)^4 - 1}{\left[(a/a_1)^4+1\right]^2}\rho_P c^2. \tag{12}$$

The pressure starts from $p = -\rho_P c^2$ at $t = -\infty$, remains approximately constant during the inflation, increases at the end of the inflation, becomes positive, reaches a maximum value p_e, and decreases algebraically during the radiation era as $p \sim (1/3)\rho c^2 \sim (1/3)\rho_P c^2 (a_1/a)^4 \sim (1/32\pi)\rho_P c^2 (t_P/t)^2$. At the transition/deceleration point $t = t_1 = t_c$, we have $p_1/(\rho_P c^2) = -1/6$. The pressure vanishes ($w = 0$) at $\rho_w/\rho_P = 1/4$, $a_w/a_1 = 3^{1/4}$. This corresponds to a time $t_w = (3/32\pi)^{1/2}\,[2-(1/2)\ln(3)-C]\,t_P$. On the other hand, the pressure reaches its maximum value $p_e/(\rho_P c^2) = 1/48$ when $\rho_e/\rho_P = 1/8$, $a_e/a_1 = 7^{1/4}$. This corresponds to a time $t_e = (3/32\pi)^{1/2}[\sqrt{8}-\ln(1+\sqrt{8})+\ln 7/2 - C]t_P$. The speed of sound is imaginary ($c_s^2 < 0$) when $a < a_e$, vanishes ($c_s = 0$) when $a = a_e$, and is real ($c_s^2 > 0$) when $a > a_e$. It is always less than the speed of light.

3.6. Evolution of the temperature

The thermodynamical equation (5) with the equation of state Eq. (7) can be integrated into

$$T = T_P\left(\frac{15}{\pi^2}\right)^{1/4}\left(1-\frac{\rho}{\rho_P}\right)^{7/4}\left(\frac{\rho}{\rho_P}\right)^{1/4} = T_P\left(\frac{15}{\pi^2}\right)^{1/4}\frac{(a/a_1)^7}{[(a/a_1)^4+1]^2}, \tag{13}$$

where T_P is a constant of integration and we have used Eq. (8) to obtain the second equality. In the radiation era $\rho \ll \rho_P$, Eq. (13) reduces to $\rho/\rho_P \sim (\pi^2/15)(T/T_P)^4$ which is the Stefan-Boltzmann law $\rho c^2 = (\pi^2/15)(k_B T)^4/\hbar^3 c^3$ provided that T_P is identified with the Planck temperature $T_P = (\hbar c^5/G k_B^2)^{1/2} = 1.42\times 10^{32}\,\mathrm{K}$. Accordingly, Eq. (13) may be viewed as a generalized Stefan-Boltzmann law [8, 12].

Using Eqs. (7), (8) and (13), we find that the entropy (6) is given by

$$S = \frac{4}{3}\left(\frac{\pi^2}{15}\right)^{1/4}\frac{\rho_P c^2}{T_P}a_1^3. \tag{14}$$

We check that it is a constant. Numerically, $S/k_B = 5.04\times 10^{87}$ [8, 12].

The temperature starts from $T = 0$ at $t = -\infty$. In the vacuum energy era, it increases exponentially rapidly with time as

$$\frac{T}{T_P} \sim \left(\frac{15}{\pi^2}\right)^{1/4}\left(1-\frac{\rho}{\rho_P}\right)^{7/4} \sim \left(\frac{15}{\pi^2}\right)^{1/4}\left(\frac{a}{a_1}\right)^7 \sim \left(\frac{15}{\pi^2}\right)^{1/4}\left(\frac{l_P}{a_1}\right)^7 e^{7(8\pi/3)^{1/2}t/t_P}. \tag{15}$$

At $t = t_i = 0$, $T_i/T_P \simeq (15/\pi^2)^{1/4}(l_P/a_1)^7$. At the transition/deceleration point $t = t_1 = t_c$, we have $T_1/T_P = (1/4)(15/\pi^2)^{1/4}$. At the point where the pressure vanishes, we have $T_w = (9/16)(5/\pi^2)^{1/4}T_P$. The temperature reaches its maximum value $T_e = (7/8)^{7/4}(15/8\pi^2)^{1/4}T_P$ at the point where the pressure reaches its maximum value. In the radiation era, the temperature decreases algebraically rapidly as

$$\frac{T}{T_P} \sim \left(\frac{15}{\pi^2}\right)^{1/4}\left(\frac{\rho}{\rho_P}\right)^{1/4} \sim \left(\frac{15}{\pi^2}\right)^{1/4}\frac{a_1}{a} \sim \left(\frac{45}{32\pi^3}\right)^{1/4}\left(\frac{t_P}{t}\right)^{1/2}. \tag{16}$$

Remark: In our model, the temperature is initially very low, increases exponentially rapidly during the inflation up to a fraction (~ 0.523) of the Planck temperature $T_P = 1.42\times 10^{32}\,\mathrm{K}$, then decreases algebraically during the radiation era. On the other hand, our model generates a value of the entropy as large as $S/k_B = 5.04\times 10^{87}$. This is very different from the standard inflationary scenario [1]. In that scenario, the universe is radiation dominated up to $t_i = 10^{-35}\,\mathrm{s}$ and expands exponentially rapidly by a factor 10^{30} in the interval $t_i < t < t_f$ with $t_f = 10^{-33}\,\mathrm{s}$. For $t > t_f$, the evolution is again radiation dominated. At $t = t_i$, the temperature is about $10^{27}\,\mathrm{K}$. This corresponds to the epoch at which most grand unified theories (GUTs) have a significant influence on the evolution of the universe. During the exponential inflation, the temperature drops drastically and one must advocate a phase of re-heating by various high energy processes (not very well understood) to restore the initial temperature.

3.7. History of the early universe

In our model, the universe "starts" at $t = -\infty$ with a vanishing radius $a = 0$, a finite density $\rho = \rho_P = 5.16 \times 10^{99}\,\mathrm{g/m^3}$, a finite pressure $p = -\rho_P c^2 = -4.64 \times 10^{116}\,\mathrm{g/m\,s^2}$, and a vanishing temperature $T = 0$. The universe exists at any time in the past and does not present any singularity. For $t < 0$, the radius of the universe is less than the Planck length $l_P = 1.62 \times 10^{-35}\,\mathrm{m}$. In the Planck era, quantum gravity should be taken into account so our semi-classical approach is probably not valid in the infinite past. We define the "original" time $t = t_i = 0$ as the time at which the radius of the universe is equal to the Planck length: $a_i = l_P = 1.62 \times 10^{-35}\,\mathrm{m}$. The corresponding density, pressure and temperature are $\rho_i \simeq \rho_P = 5.16 \times 10^{99}\,\mathrm{g/m^3}$, $p_i \simeq -\rho_P c^2 = -4.64 \times 10^{116}\,\mathrm{g/m\,s^2}$ and $T_i = 3.91 \times 10^{-205}\,T_P = 5.54 \times 10^{-173}\,\mathrm{K}$. We note that quantum mechanics regularizes the finite time singularity present in the standard Big Bang theory. This is similar to finite size effects in second order phase transitions (the standard Big Bang theory is recovered for $\hbar \to 0$) [8, 12]. We also note that the universe is very cold at $t = t_i = 0$, unlike what is predicted by the Big Bang theory (a naive extrapolation of the law $T \propto t^{-1/2}$ leads to $T(0) = +\infty$). The universe first undergoes a phase of inflation driven by the vacuum energy during which its radius and temperature increase exponentially rapidly while its density and pressure remain approximately constant. The inflation "starts" at $t_i = 0$ and "ends" at $t_1 = 23.3\,t_P = 1.25 \times 10^{-42}\,\mathrm{s}$. During this very short lapse of time, the radius of the universe grows from $a_i = l_P = 1.62 \times 10^{-35}\,\mathrm{m}$ to $a_1 = 2.61 \times 10^{-6}\,\mathrm{m}$, and the temperature grows from $T_i = 3.91 \times 10^{-205}\,T_P = 5.54 \times 10^{-173}\,\mathrm{K}$ to $T_1 = 0.278\,T_P = 3.93 \times 10^{31}\,\mathrm{K}$. By contrast, the density and the pressure do not change significatively: they go from $\rho_i \simeq \rho_P = 5.16 \times 10^{99}\,\mathrm{g/m^3}$ and $p_i \simeq -\rho_P c^2 = -4.64 \times 10^{116}\,\mathrm{g/m\,s^2}$ to $\rho_1 = \rho_P/2 = 2.58 \times 10^{99}\,\mathrm{g/m^3}$ and $p_1 = -(1/6)\rho_P c^2 = -7.73 \times 10^{115}\,\mathrm{g/m\,s^2}$. The pressure passes from negative values to positive values at $t_w = 23.4 t_P = 1.26 \times 10^{-42}\,\mathrm{s}$ (at that moment $\rho_w = \rho_P/4 = 1.29 \times 10^{99}\,\mathrm{g/m^3}$, $a_w = 1.32\,a_1 = 3.43 \times 10^{-6}\,\mathrm{m}$ and $T_w = 0.475 T_P = 6.74 \times 10^{31}\,\mathrm{K}$). After the inflation, the universe enters in the radiation era and, from that point, we recover the standard model [2]. The radius increases algebraically as $a \propto t^{1/2}$ while the density and the temperature decrease algebraically as $\rho \propto t^{-2}$ and $T \propto t^{-1/2}$. The pressure and the temperature achieve their maximum values $p_e = 2.08 \times 10^{-2} \rho_P c^2 = 9.66 \times 10^{114}\,\mathrm{g/m\,s^2}$ and $T_e = 0.523\,T_P = 7.40 \times 10^{31}\,\mathrm{K}$ at $t = t_e = 23.6\,t_P = 1.27 \times 10^{-42}\,\mathrm{s}$. At that moment, the density is $\rho_e = \rho_P/8 = 6.44 \times 10^{98}\,\mathrm{g/m^3}$ and the radius $a_e = 1.63\,a_1 = 4.24 \times 10^{-6}\,\mathrm{m}$. During the inflation, the universe is accelerating and during the radiation era it is decelerating. The transition (marked by an inflexion point) takes place at a time $t_c = t_1 = 23.3\,t_P = 1.25 \times 10^{-42}\,\mathrm{s}$ coinciding with the end of the inflation ($a_c = a_1$). The evolution of the scale factor, density and temperature of the universe as a function of time are represented in Figs. 1-3 below.

3.8. Scalar field: inflaton

The phase of inflation in the very early universe is usually described by some hypothetical scalar field ϕ, called inflaton, with its origin in the quantum fluctuations of the vacuum [1]. A canonical scalar field minimally coupled to gravity evolves according to the Klein-Gordon equation

$$\ddot{\phi} + 3H\dot{\phi} + \frac{dV}{d\phi} = 0, \tag{17}$$

where $V(\phi)$ is the potential of the scalar field. The scalar field tends to run down the potential towards lower energies. The density and the pressure of the universe are related to the scalar field by

$$\rho c^2 = \frac{1}{2}\dot{\phi}^2 + V(\phi), \qquad p = \frac{1}{2}\dot{\phi}^2 - V(\phi). \tag{18}$$

Using standard techniques [9, 12], we find that the inflaton potential corresponding to the equation of state (7) is

$$V(\psi) = \frac{1}{3}\rho_P c^2 \frac{\cosh^2\psi + 2}{\cosh^4\psi} \qquad \psi = \left(\frac{8\pi G}{c^2}\right)^{1/2}\phi, \qquad (\phi \geq 0). \tag{19}$$

We can also show [9, 12] that the scale factor, the energy density and the pressure of the universe are related to the scalar field by

$$\left(\frac{a}{a_1}\right)^2 = \sinh\psi, \qquad \rho = \frac{\rho_P}{\cosh^2\psi}, \qquad p = \frac{\rho_P c^2}{3\cosh^2\psi}\left(1 + \frac{2}{\cosh^2\psi}\right). \tag{20}$$

Using Eq. (20) and the results of the previous sections we can obtain the temporal evolution of the inflaton in the early universe [9, 12].

For $\psi \to 0$, the scalar field potential (19) can be expanded up to fourth order giving

$$\frac{V(\psi)}{\rho_P c^2} \simeq 1 - \frac{5}{3}\psi^2 + \frac{16}{9}\psi^4 + \ldots \tag{21}$$

This expression can be compared with the standard form of a quartic potential

$$V = V_0 + \frac{m^2 c^4}{2\hbar^2}\phi^2 + \frac{\lambda c^3}{4\hbar}\phi^4, \tag{22}$$

where V_0 is the energy of point zero, m is the mass of the SF and λ is the dimensionless self-interaction constant. When the SF describes the wave function of a Bose-Einstein condensate (BEC), we have

$$\frac{\lambda}{8\pi} = \frac{a_s}{\lambda_C} = \frac{a_s mc}{\hbar}, \tag{23}$$

where a_s is the scattering length of the bosons and λ_C is their Compton length. Comparing Eq. (21) with Eq. (22), we obtain

$$V_0 = \rho_P c^2, \tag{24}$$

$$m^2 = -\frac{80\pi}{3}\frac{\rho_P G\hbar^2}{c^4} = -\frac{80\pi}{3}M_P^2, \tag{25}$$

$$\lambda = \frac{4096\pi^2}{9}, \qquad a_s^2 = -\frac{16384\pi}{135}l_P^2, \tag{26}$$

where $M_P = (\hbar c/G)^{1/2} = 2.17 \times 10^{-5}$ g is the Planck mass. From Eqs. (25) and (26) we get

$$a_s = \frac{16}{15}\frac{2Gm}{c^2} = \frac{16}{15}r_S, \tag{27}$$

where $r_S = 2Gm/c^2$ is the effective Schwarzschild radius of the particle. We note that the mass of the inflaton is imaginary, but its modulus is of the order of the Planck mass (similarly $\lambda \sim 1$ and $|a_s| \sim l_P$). In theories of extended supergravity, the mass is quantized according to the rule [13]:

$$m^2 = n_* \frac{H^2 \hbar}{c^4}, \tag{28}$$

where n_* is an integer and H is the Hubble factor in the de Sitter era. Since $H^2 = 8\pi G\rho_P/3$ during the early inflation, we can rewrite Eq. (28) as

$$m^2 = n_* \frac{8\pi G\rho_P \hbar}{3c^4}. \tag{29}$$

We see that this quantization condition is consistent with Eq. (25) for the integer value $n_* = -10$ [14].

4. Late universe: dark matter + dark energy

4.1. Equation of state

We now consider the equation of state (1) with $\alpha = 0$, $n = -1$ and $k = -\rho_\Lambda$ which provides a model of late universe dominated by dark matter and dark energy. This equation of state can be rewritten as [9, 12]:

$$p = -\rho_\Lambda c^2. \tag{30}$$

The pressure has a constant negative value. The continuity equation (2) can be integrated into

$$\rho = \rho_\Lambda \left[\left(\frac{a_2}{a} \right)^3 + 1 \right], \tag{31}$$

where a_2 is a constant of integration. The relation (31) between the energy density and the scale factor is formally equivalent to that of the ΛCDM model.

The equation of state parameter, the deceleration parameter, and the speed of sound are given by

$$w = -\frac{\rho_\Lambda}{\rho}, \qquad q = \frac{1}{2} - \frac{3}{2}\frac{\rho_\Lambda}{\rho}, \qquad \frac{c_s^2}{c^2} = 0. \tag{32}$$

As the universe expands from $a = 0$ to $a = +\infty$, the density decreases from $+\infty$ to ρ_Λ, the equation of state parameter w decreases from 0 to -1, the deceleration parameter q decreases from 1/2 to -1, and the speed of sound c_s remains equal to zero.

4.2. Dark energy era: late inflation

When $a \gg a_2$, the density is approximately constant: $\rho \simeq \rho_\Lambda$. Since this solution is expected to describe the very late universe, it is natural to identify the constant ρ_Λ with the cosmological density $\rho_\Lambda = \Lambda/8\pi G = 7.02 \times 10^{-24}\,\mathrm{g/m^3}$ (dark energy) which may represent a fundamental lower bound for the density.[1] A constant value of the density gives rise to a phase of late inflation (accelerated expansion of the universe). It is convenient to define a cosmological time $t_\Lambda = 1/(G\rho_\Lambda)^{1/2} = (8\pi/\Lambda)^{1/2} = 1.46 \times 10^{18}\,\mathrm{s}$, a cosmological length $l_\Lambda = ct_\Lambda = (8\pi c^2/\Lambda)^{1/2} = 4.38 \times 10^{26}\,\mathrm{m}$ and a cosmological mass $M_\Lambda = \rho_\Lambda l_\Lambda^3 = (8\pi c^6/\Lambda G^2)^{1/2} = 5.90 \times 10^{56}\,\mathrm{g}$. These are the counterparts of the Planck scales for the late universe [9]. Since $\Omega_{\Lambda,0} \sim 1$, these cosmological scales are of the same order as the Hubble scales defined in terms of $\rho_0 = 3H_0^2/8\pi G$. From the Friedmann equation (3), we find that the scale factor increases exponentially rapidly with time as

$$a(t) \sim l_\Lambda e^{(8\pi/3)^{1/2}(t-t_f)/t_\Lambda}, \tag{33}$$

where t_f (determined below) is the time at which $a_f = l_\Lambda$ and $\rho_f = [(a_2/l_\Lambda)^3 + 1]\rho_\Lambda$. This exponential growth corresponds to the de Sitter (dS) solution. The timescale of the exponential growth is the cosmological time $t_\Lambda = 1.46 \times 10^{18}\mathrm{s}$. This solution exists at any time in the future ($a \to +\infty$ and $\rho \to \rho_\Lambda$ for $t \to +\infty$), so there is no future singularity.[2]

[1] In our model, ρ_Λ and Λ appear as fundamental constants that characterize the late universe (in the same sense that ρ_P and $\hbar$ characterize the early universe). To determine the value of ρ_Λ, we use the results of observations and the fact that our model of late universe is formally equivalent to the ΛCDM model. Thus, $\rho_\Lambda = \Omega_{\Lambda,0}\rho_0$ with $\Omega_{\Lambda,0} = 0.763$ and $\rho_0 = 3H_0^2/8\pi G = 9.20 \times 10^{-24}\mathrm{g/m^3}$ giving $\rho_\Lambda = 7.02 \times 10^{-24}\,\mathrm{g/m^3}$.

[2] This is not the case of all cosmological models. In a phantom universe [11], violating the null dominant energy condition ($w < -1$), the density increases as the universe expands. The models based on phantom dark energy usually predict a future singularity, called Big Rip [15], in which the scale factor and the energy density become infinite in a finite time. There can also be a Little Rip [16] in which the scale factor and the energy density becomes infinite in infinite time. Contrary to the Big Crunch, the universe is destroyed not by excessive contraction but rather by excessive expansion. The possibility that we live in a phantom universe is not ruled out by observations.

4.3. Matter era

When $a \ll a_2$, we get $\rho/\rho_\Lambda \sim (a_2/a)^3$. Because of the smallness of ρ_Λ, the equation of state (30) can be approximated by $p \sim 0$ which describes the pressureless matter era.[3] The conservation of $\rho_{\rm m}\, a^3$ implies that $\rho_\Lambda a_2^3 = \rho_{\rm m,0}\, a_0^3$. Using $\rho_\Lambda = \Omega_{\Lambda,0}\rho_0$ and $\rho_{\rm m,0} = \Omega_{\rm m,0}\rho_0$ with $\Omega_{\Lambda,0} = 0.763$ (dark energy) and $\Omega_{\rm m,0} = \Omega_{\rm b,0} + \Omega_{\rm dm,0} = 0.237$ (baryonic and dark matter), we obtain $a_2/a_0 = (\Omega_{m,0}/\Omega_{\Lambda,0})^{1/3} = 0.677$, hence $a_2 = 8.95 \times 10^{25}\,{\rm m}$. This can be rewritten as $a_2 = 0.204 l_\Lambda$. The characteristic scale a_2 marks the transition between the matter era and the dark energy era. When $a \ll a_2$, the Friedmann equation (3) yields $a/a_2 \sim (6\pi)^{1/3}(t/t_\Lambda)^{2/3}$ and $\rho/\rho_\Lambda \sim (1/6\pi)(t_\Lambda/t)^2$. This is the Einstein-de Sitter (EdS) solution of a pressureless universe without cosmological constant.

4.4. The general solution

The general solution of the Friedmann equation (3) with Eq. (31) is [9, 12]:

$$\frac{a}{a_2} = \sinh^{2/3}\left(\sqrt{6\pi}\frac{t}{t_\Lambda}\right), \qquad \frac{\rho}{\rho_\Lambda} = \frac{1}{\tanh^2\left(\sqrt{6\pi}\frac{t}{t_\Lambda}\right)}. \tag{34}$$

Its asymptotic behavior for $t \to +\infty$ returns Eq. (33) with $t_f = (3/8\pi)^{1/2}[(2/3)\ln 2 + \ln(l_\Lambda/a_2)]t_\Lambda$. The universe is decelerating when $a < a'_c$ (i.e. $\rho > \rho'_c$) and accelerating when $a > a'_c$ (i.e. $\rho < \rho'_c$) where $a'_c = (1/2)^{1/3}a_2$ and $\rho'_c = 3\rho_\Lambda$. The time t'_c at which the universe starts accelerating is given by $t'_c = (1/6\pi)^{1/2}{\rm argsinh}(1/\sqrt{2})t_\Lambda$. This corresponds to the time at which the curve $a(t)$ presents a second inflexion point. The time t_2 corresponding to the transition between the matter era and the dark energy era ($a = a_2$) is $t_2 = (1/\sqrt{6\pi}){\rm argsinh}(1)t_\Lambda$. The corresponding density is $\rho_2 = 2\rho_\Lambda$.

4.5. The temperature of radiation

In the late universe, the temperature of radiation [see Eq. (16)] is given by

$$\frac{T}{T_P} = \left(\frac{15}{\pi^2}\right)^{1/4}\frac{a_1}{a}, \tag{35}$$

where the scale factor is given by Eq. (34). In the matter era

$$\frac{T}{T_P} = \left(\frac{15}{\pi^2}\right)^{1/4}\frac{a_1}{a_2}\frac{1}{(6\pi)^{1/3}}\left(\frac{t_\Lambda}{t}\right)^{2/3}. \tag{36}$$

In the dark energy era

$$\frac{T}{T_P} = \left(\frac{15}{\pi^2}\right)^{1/4}\frac{a_1}{l_\Lambda}e^{-(8\pi/3)^{1/2}(t-t_f)/t_\Lambda}. \tag{37}$$

4.6. History of the late universe

When $a \ll a_2$, the universe is in the matter era. Its radius increases algebraically as $a \propto t^{2/3}$ while its density decreases algebraically as $\rho \propto t^{-2}$ (EdS). The temperature of radiation decreases algebraically as $T \propto t^{-2/3}$. The expansion of the universe is decelerating. When $a \gg a_2$, the universe is in the dark energy era. It undergoes a phase of accelerating expansion (late inflation) driven by the dark energy during which its radius increases exponentially rapidly (dS) while

[3] The equation of state (30) applies only to dark matter and dark energy. In principle, baryonic matter should be treated as a different fluid with $p_{\rm b} = 0$ (see Sec. 5). However, for commodity, we shall include the contribution of baryonic matter in the density ρ because the final equations are the same.

its density remains constant and equal to the cosmological density $\rho_\Lambda = 7.02 \times 10^{-24}\,\mathrm{g/m^3}$. The temperature of radiation decreases exponentially rapidly. The time at which the universe starts accelerating is $t'_c = 0.152 t_\Lambda = 2.22 \times 10^{17}\,\mathrm{s}$, corresponding to a radius $a'_c = 0.794 a_2 = 0.162 l_\Lambda = 7.11 \times 10^{25}\mathrm{m}$ and a density $\rho'_c = 3\rho_\Lambda = 2.11 \times 10^{-23}\,\mathrm{g/m^3}$. The transition between the matter era and the dark energy era takes place at $a_2 = 0.204 l_\Lambda = 8.95 \times 10^{25}\,\mathrm{m}$, $\rho_2 = 2\rho_\Lambda = 1.40 \times 10^{-23}\,\mathrm{g/m^3}$, $t_2 = 0.203 t_\Lambda = 2.97 \times 10^{17}\,\mathrm{s}$. At $t = t_f = 0.708 t_\Lambda = 1.03 \times 10^{18}\,\mathrm{s}$, the radius of the universe is equal to the cosmological length ($a_f = l_\Lambda = 4.38 \times 10^{26}\,\mathrm{m}$) and its density $\rho_f = 1.008\rho_\Lambda = 7.08 \times 10^{-24}\mathrm{g/m^3}$. The evolution of the scale factor, density and temperature of the universe as a function of time are represented in Figs. 1-3 below. The pressure has a constant negative value $p = -\rho_\Lambda c^2 = -6.31 \times 10^{-7}\,\mathrm{g/ms^2}$.

4.7. The present universe

The present size of the visible universe $a_0 = 0.302 l_\Lambda = 1.32 \times 10^{26}\,\mathrm{m}$ is precisely of the order of the scale $a_2 = 0.204 l_\Lambda = 8.95 \times 10^{25}\,\mathrm{m}$ ($a_0 = 1.48 a_2$). Therefore, we live just at the transition between the matter era and the dark energy era (see the bullets in Figs. 1-3). The present density of the universe is $\rho_0 = 1.31\rho_\Lambda = 9.20 \times 10^{-24}\,\mathrm{g/m^3}$. The present values of the deceleration parameter and of the equation of state parameter are $q_0 = (1 - 3\Omega_{\Lambda,0})/2 = -0.645$ and $w_0 = -\Omega_{\Lambda,0} = -0.763$. The age of the universe is $t_0 = (1/6\pi)^{1/2}\mathrm{argsinh}[(a_0/a_2)^{3/2}]t_\Lambda$. Numerically, $t_0 = 0.310 t_\Lambda = 4.54 \times 10^{17}\,\mathrm{s} \sim 14\,\mathrm{Gyrs}$ ($t_0 = 13.7\,\mathrm{Gyrs}$ if we use a more precise value of H_0). We have $t_0 = 1.03\,H_0^{-1}$. It is rather fortunate that the age of the universe almost coincides with the Hubble time H_0^{-1}. The present temperature of radiation is $T_0 = T_P(15/\pi^2)^{1/4}(a_1/a_0)$. Numerically, $T_0 = 3.11\,\mathrm{K}$ ($T_0 = 2.7\,\mathrm{K}$ if we take into account the radiation of neutrinos in thermal equilibrium).

4.8. Scalar field: quintessence

In alternative theories to the cosmological constant, the dark energy is usually described by a scalar field, defined by Eqs. (17) and (18), called quintessence [4, 3]. Using standard techniques [9, 12], we find that the quintessence potential corresponding to the equation of state (30) is

$$V(\psi) = \frac{1}{2}\rho_\Lambda c^2 \left(\cosh^2\psi + 1\right), \qquad \psi = \frac{3}{2}\left(\frac{8\pi G}{3c^2}\right)^{1/2}(\phi - \phi_{\max}) \qquad (\psi \leq 0). \tag{38}$$

We can also show [9, 12] that the scale factor and the energy density of the universe are related to the scalar field by

$$\left(\frac{a_2}{a}\right)^{3/2} = -\sinh\psi, \qquad \rho = \rho_\Lambda \cosh^2\psi. \tag{39}$$

Using Eq. (39) and the results of the previous sections we can obtain the temporal evolution of the quintessence field in the late universe [9, 12].

For $\psi \to 0$, the scalar field potential (38) can be expanded up to fourth order giving

$$\frac{V(\psi)}{\rho_\Lambda c^2} \simeq 1 + \frac{1}{2}\psi^2 + \frac{1}{6}\psi^4 + ... \tag{40}$$

Comparing this expression with the standard form (22) of a quartic potential, we obtain

$$V_0 = \rho_\Lambda c^2, \tag{41}$$

$$m^2 = \frac{6\pi G \rho_\Lambda \hbar}{c^4}, \tag{42}$$

$$\lambda = 24\pi^2 \frac{\rho_\Lambda}{\rho_P}, \qquad a_s^2 = \frac{3\pi G^3 \rho_\Lambda \hbar}{2c^8}. \tag{43}$$

IOP Conf. Series: Journal of Physics: Conf. Series **1030** (2018) 012009 doi:10.1088/1742-6596/1030/1/012009

From Eqs. (42) and (43) we get

$$a_s = \frac{Gm}{2c^2} = \frac{1}{4} r_S. \tag{44}$$

We note that the mass of the quintessence field is real and is of order $m_\Lambda = \hbar H_0/c^2 = 1.43 \times 10^{-33}\,\mathrm{eV}/c^2$ (since $H_0^2 \sim G\rho_\Lambda \sim \Lambda$). This mass scale is often interpreted as the smallest mass of the bosons predicted by string theory [17] or as the upper bound on the mass of the graviton [18]. It is simply obtained by equating the Compton wavelength of the particle $\lambda_C = \hbar/mc$ with the Hubble radius $a_0 = c/H_0$ (the typical size of the visible universe). On the other hand, the self-interaction constant λ and the scattering length a_s are of the order of $\rho_\Lambda/\rho_P \sim 10^{-123}$ and $a_\Lambda = G\hbar H_0/c^4 = 1.90 \times 10^{-81}$ fm.

Since $H^2 = 8\pi G\rho_\Lambda/3$ during the late inflation, we can rewrite the quantization condition (28) as

$$m^2 = n_* \frac{8\pi G\rho_\Lambda \hbar}{3c^4}. \tag{45}$$

We see that Eq. (42) is *not* consistent with this quantization condition since it corresponds to $n_* = 9/4$ which is not an integer. We recall that our model of late universe is formally equivalent to the ΛCDM model. As a result, the quintessence field associated with the standard ΛCDM model does not satisfy the quantization condition (28) [14].

Remark: In this paper, we have described the late universe by a polytrope of index $n = -1$. We have shown that it is equivalent to the ΛCDM model. If we consider a polytropic equation of state with an arbitrary index $n < 0$, we find that the mass of the scalar field is given by [14]:

$$m^2 = -\frac{1+2n}{n^2}\frac{6\pi G\rho_\Lambda \hbar}{c^4}. \tag{46}$$

We note that this mass presents a maximum for $n = -1$. This maximum selects the ΛCDM model among all the polytropic models of the form of Eq. (1) with $n < 0$ [14].

4.9. Scalar field: tachyons

Some authors have proposed to represent the dark energy by a rolling tachyon condensate appearing in a class of string theories [19]. A tachyonic scalar field [20, 21, 22] has an equation of state $p = w(t)\rho c^2$ with $-1 \le w(t) \le 0$. This scalar field evolves according to the equation

$$\frac{\ddot\phi}{1-\dot\phi^2} + 3H\dot\phi + \frac{1}{V}\frac{dV}{d\phi} = 0. \tag{47}$$

The tachyonic field tends to run down the potential towards lower energies. The density and the pressure of the universe are related to the tachyonic field by

$$\rho c^2 = \frac{V(\phi)}{\sqrt{1-\dot\phi^2}}, \qquad p = -V(\phi)\sqrt{1-\dot\phi^2}. \tag{48}$$

Since $w = p/\rho c^2 = -\rho_\Lambda/\rho$ is between -1 and 0, the equation of state (30) can be associated with a tachyonic field. Using standard techniques [9, 12], we find that its potential is

$$V(\psi) = \frac{\rho_\Lambda c^2}{\cos\psi}, \qquad \psi = \sqrt{\rho_\Lambda c^2}\left(\frac{6\pi G}{c^2}\right)^{1/2}(\phi - \phi_{\rm max}) \qquad (-\pi/2 \le \psi \le 0). \tag{49}$$

We can also show [9, 12] that the scale factor and the density of the universe are related to the scalar field by

$$\left(\frac{a_2}{a}\right)^{3/2} = -\tan\psi, \qquad \rho = \frac{\rho_\Lambda}{\cos^2\psi}. \tag{50}$$

Using Eq. (50) and the results of the previous sections we can obtain the temporal evolution of the tachyonic scalar field in the late universe [9, 12].

5. The whole evolution of the universe

5.1. *First point of view: Generalized radiation + dark fluid*

In Sec. 3, we have described the transition between the vacuum energy era (Planck era) and the radiation era in the early universe by a single equation of state (7). This equation of state interpolates smoothly between the equation of state of vacuum energy ($p = -\rho c^2$) and the equation of state of radiation ($p = \rho c^2/3$). It provides therefore a unified description of vacuum energy and radiation. We shall consider that it describes a fluid of "generalized radiation". The relation between the energy density of this fluid and the scale factor can be written as

$$\frac{\rho}{\rho_0} = \frac{\Omega_{\rm rad,0}}{\frac{\Omega_{\rm rad,0}\rho_0}{\rho_P} + (a/a_0)^4}, \tag{51}$$

where $\Omega_{\rm rad,0}\rho_0/\rho_P = 1.51 \times 10^{-127}$.

In Sec. 4, we have described the transition between the dark matter era and the dark energy era (de Sitter era) in the late universe by a single equation of state (30). This equation of state interpolates smoothly between the equation of state of pressureless matter ($p = 0$) and the equation of state of dark energy ($p = -\rho c^2$). It provides therefore a unified description of dark matter and dark energy. We shall consider that it describes a "dark fluid". The relation between the energy density of this fluid and the scale factor can be written as

$$\frac{\rho}{\rho_0} = \frac{\Omega_{\rm m,0}}{(a/a_0)^3} + \Omega_{\Lambda,0}. \tag{52}$$

This is formally equivalent to the ΛCDM model.

We can combine these two models by assuming the the universe is filled with a fluid of "generalized radiation" (vacuum energy + radiation) and a "dark fluid" (dark matter + dark energy). Summing the densities of these two fluids, considered as two different species, we get

$$\frac{\rho}{\rho_0} = \frac{\Omega_{\rm rad,0}}{\frac{\Omega_{\rm rad,0}\rho_0}{\rho_P} + (a/a_0)^4} + \frac{\Omega_{\rm m,0}}{(a/a_0)^3} + \Omega_{\Lambda,0}. \tag{53}$$

5.2. *Second point of view: Unitary fluid + dark matter*

Alternatively, we can unify vacuum energy, radiation and dark energy by a single (quadratic) equation of state of the form [23, 24]:[4]

$$p = -\frac{4\rho^2}{3\rho_P}c^2 + \frac{1}{3}\rho c^2 - \frac{4}{3}\rho_\Lambda c^2. \tag{54}$$

It is interesting to note that the Planck density ρ_P, the radiation parameter $\alpha = 1/3$ and the cosmological density ρ_Λ appear as the coefficients of this quadratic equation of state. We shall consider that this equation of state describes a "unitary fluid". We can integrate analytically the energy equation (2) with the equation of state (54) without approximation [24]. However, using the fact that $\rho_P \gg \rho_\Lambda$, we obtain in very good approximation [23, 24]:

$$\frac{\rho}{\rho_0} = \frac{\Omega_{\rm rad,0}}{\frac{\Omega_{\rm rad,0}\rho_0}{\rho_P} + (a/a_0)^4} + \Omega_{\Lambda,0}. \tag{55}$$

[4] Many cosmological models, such as the Chaplygin gas model [5], attempt to unify two fluids: dark matter and dark energy. The quadratic equation of state (54) attempts to unify three fluids: vacuum energy, radiation and dark energy. This model is discussed in detail in [23, 24]. In particular, we determine the scalar field potential corresponding to the equation of state (54) that unifies the inflaton and quintessence fields.

IOP Conf. Series: Journal of Physics: Conf. Series **1030** (2018) 012009 doi:10.1088/1742-6596/1030/1/012009

On the other hand, we consider dark matter as an independant fluid with a pressureless equation of state $p = 0$. In that case, the energy equation (2) gives

$$\frac{\rho}{\rho_0} = \frac{\Omega_{\rm m,0}}{(a/a_0)^3}. \tag{56}$$

Assuming that the universe is filled with a "unitary fluid" (vacuum energy + radiation + dark energy) and "dark matter", and summing the densities of these two fluids, we obtain again Eq. (53).

5.3. A cosmic equation that generalizes the ΛCDM model

Substituting Eq. (53) into the Friedmann equation (3) which can be rewritten as $(H/H_0)^2 = \rho/\rho_0$, we obtain a cosmic equation of the form [9, 12, 23, 24]:

$$\left(\frac{H}{H_0}\right)^2 = \frac{\Omega_{\rm rad,0}}{\frac{\Omega_{\rm rad,0}\rho_0}{\rho_P} + (a/a_0)^4} + \frac{\Omega_{\rm m,0}}{(a/a_0)^3} + \Omega_{\Lambda,0}. \tag{57}$$

This equation describes the whole evolution of the universe, from its early inflation to its late acceleration (see Figs. 1-3). For $\hbar \to 0$ (i.e. $\rho_P \to +\infty$), we recover the standard ΛCDM model which presents a singularity at $t = 0$ (Big Bang) and does not describe the phase of early inflation. For $\hbar \neq 0$, we obtain a model of universe in which the initial singularity is removed and the universe displays a phase of early inflation. After the early inflation, i.e. for $t > t_1 = 23.3t_P = 1.25 \times 10^{-42}$ s, this model gives the same results as the ΛCDM model: the universe undergoes a decelerating power-law expansion in the radiation and matter eras, then an accelerated exponential expansion (late inflation) in the dark energy era. A nice feature of this model is its simplicity since it incorporates a phase of early inflation in a very simple manner. We just have to add a cut-off $(a_1/a_0)^4 = \Omega_{\rm rad,0}\rho_0/\rho_P$ in the cosmic equation of the standard ΛCDM model. Therefore, the modification implied by Eq. (57) to describe the early inflation is very natural. On the other hand, the equation of state (30) in the late universe gives the same results as the ΛCDM model where we add the contribution of dark matter and dark energy individually, so this does not bring any modification to the standard cosmic equation. This is an interest of our description since the ΛCDM model gives results that agree with observations at late times. In our model, the universe exists eternally in the past[5] and in the future without any singularity (aioniotic universe). As explained in Ref. [9], this model is the most natural non-singular symmetric solution of the cosmological Einstein equations.

5.4. Cosmological constant problem

It is oftentimes argued that the dark energy ρ_Λ (cosmological constant) corresponds to the vacuum energy. Since the vacuum energy density is expected to be of the order of the Planck density ρ_P this leads to the so-called cosmological constant problem [26, 27] because the cosmological density $\rho_\Lambda = 7.02 \times 10^{-24}\,{\rm g/m^3}$ and the Planck density $\rho_P = 5.16 \times 10^{99}\,{\rm g/m^3}$ differ by about 123 orders of magnitude ($\rho_P/\rho_\Lambda \sim 10^{123}$). We think it is a mistake to identify the dark energy with the vacuum energy. As illustrated in Fig. 2, the Planck density and the cosmological density represent fundamental upper and lower density bounds acting in the early

[5] As already stated in Sec. 3, our model certainly breaks down in the Planck era ($t \leq 0$) where $a \leq l_P$. The Planck era may not be described in terms of an equation of state $p(\rho)$, or even in terms of the Einstein equations, as we have assumed. It probably requires the development of a theory of quantum gravity. An interesting description of the early inflation has been given in Ref. [25] in terms of a quantized model based on a simplified Wheeler-DeWitt equation. In that model, a quantum tunneling process explains the birth of the universe with a well defined size after tunneling.

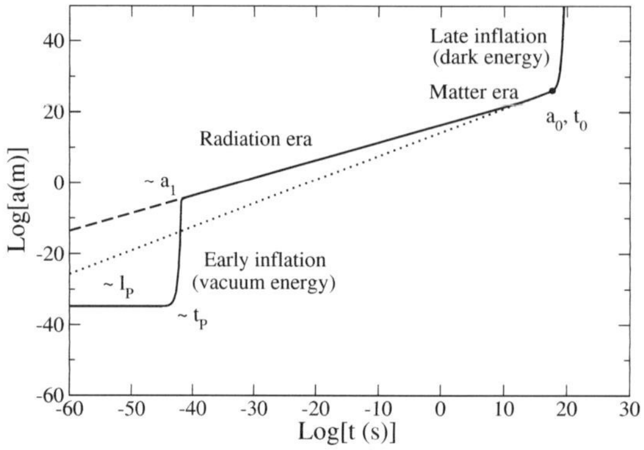

Figure 1. Evolution of the scale factor a as a function of time obtained by integrating Eq. (57). The universe undergoes an early inflation driven by the Planck density ρ_P connecting the vacuum energy era (Planck era) to the radiation era. During the early inflation, the scale factor increases by 29 orders of magnitude in less than 10^{-42} s. In the radiation era, the scale factor increases algebraically as $t^{1/2}$ and in the matter era it increases algebraically as $t^{2/3}$. The universe finally undergoes a late inflation (accelerated expansion) driven by the cosmological density ρ_Λ connecting the matter era to the dark energy era (de Sitter era). At present, we live just at the transition (bullet). The universe is accelerating during the vacuum energy era, decelerating during the radiation and matter eras, and accelerating again during the dark energy era. The evolution of the early and late universe is remarkably symmetric. In our model, it is described by two polytropic equations of state with index $n = 1$ and $n = -1$ respectively. The dashed and dotted lines corresponds to the ΛCDM model (with or without radiation) leading to a primordial singularity at $t = 0$ (Big Bang).

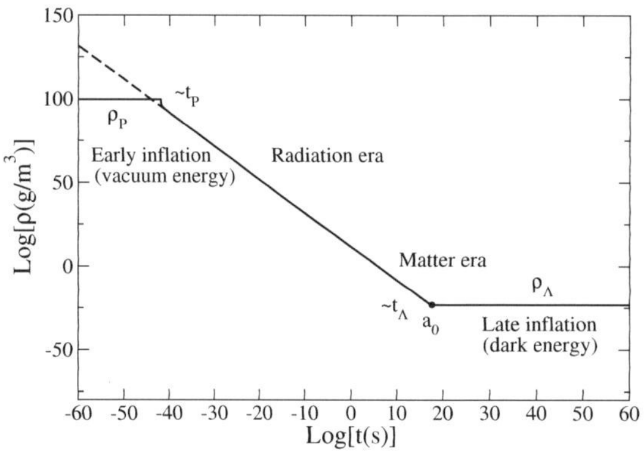

Figure 2. Evolution of the energy density ρ as a function of time from Eq. (57). The density goes from a maximum value $\rho_{\max} = \rho_P$ determined by the Planck constant (quantum mechanics) to a minimum value $\rho_{\min} = \rho_\Lambda$ determined by the cosmological constant (general relativity). These two bounds, which are fixed by fundamental constants of physics, are responsible for the early and late inflation (acceleration) of the universe. They differ by a factor of the order 10^{123}. In the radiation and matter eras, the density decreases algebraically as t^{-2}.

and late universe, respectively. Therefore, it is not surprising that they differ by 123 orders of magnitude. In this paper, we regard the vacuum energy and the dark energy as two distinct

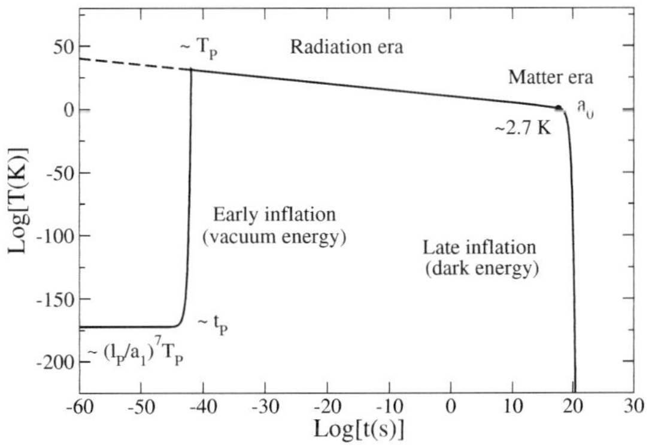

Figure 3. Evolution of the temperature of radiation as a function of time ($T/T_0 = (a/a_0)^7/[(a/a_0)^4 + (a_1/a_0)^4]^2$). Before the early inflation, the universe is extremely cold ($T < 10^{-173}$ K). During the early inflation, the temperature increases by 204 orders of magnitude in less than 10^{-42} s. During the radiation and matter eras, the temperature decreases algebraically as $T \sim t^{-1/2}$ and $T \sim t^{-2/3}$, respectively. The present value of the temperature is $T_0 \simeq 2.7$ K (bullet). During the late inflation, the temperature decreases exponentially rapidly.

entities. We call vacuum energy the energy associated with the Planck density ρ_P and dark energy the energy associated with the cosmological density ρ_Λ. The vacuum energy is responsible for the inflation of the early universe and the dark energy for the inflation (acceleration) of the late universe. In this viewpoint, the vacuum energy is due to quantum mechanics and the dark energy is an effect of general relativity. The cosmological constant Λ may be interpreted as a fundamental constant of nature applying to the cosmophysics (late universe) in the same way the Planck constant $\hbar$ applies to the microphysics (early universe). Accordingly, the origin of the dark energy density ρ_Λ should not be sought in quantum mechanics, but in pure general relativity. In this sense, the cosmological constant "problem" may be a false problem. Of course, the origin of the cosmological constant (or dark energy) still remains to be understood.

5.5. Further generalisations of the cosmic equation (57)

We have described the late universe by a constant equation of state $p/c^2 = -\rho_\Lambda$ interpolating between pressureless matter ($p = 0$) and dark energy ($p = -\rho c^2$). More generally, we could consider an equation of state of the form $p/c^2 = \alpha\rho - (\alpha + 1)\rho_\Lambda$ [8, 9] interpolating between isothermal matter ($p = \alpha\rho c^2$) and dark energy ($p = -\rho c^2$). In that case, Eq. (57) is replaced by

$$\left(\frac{H}{H_0}\right)^2 = \frac{\Omega_{\rm rad,0}}{\frac{\Omega_{\rm rad,0}\rho_0}{\rho_P} + (a/a_0)^4} + \frac{\Omega_{\rm m,0}}{(a/a_0)^{3(1+\alpha)}} + \Omega_{\Lambda,0}. \tag{58}$$

We have assumed that the universe is made of vacuum energy, radiation, dark matter and dark energy (some of them being unified in one fluid). We could also add the contribution of other species such as baryonic matter, stiff matter [28, 29, 30], or more generally an hypothetical exotic α-fluid [24]. Finally, we could add the contribution of a scalar field representing for example BEC dark matter [31, 32, 33, 34].

We have described the early universe by a polytrope of index $n = +1$ and the late universe by a polytrope of index $n = -1$. More generally, we could describe the early universe by a polytrope of index $n_e > 0$ and the late universe by a polytrope of index $n_l < 0$ [8, 9, 24].

IOP Conf. Series: Journal of Physics: Conf. Series **1030** (2018) 012009 doi:10.1088/1742-6596/1030/1/012009

A cosmic equation taking into account some of the possible generalizations proposed previously is given by [24]:

$$\left(\frac{H}{H_0}\right)^2=\frac{\Omega_{\alpha,0}}{\left[(a/a_0)^{3(1+\alpha)/n_e}+(a_1/a_0)^{3(1+\alpha)/n_e}\right]^{n_e}}+\frac{\Omega_{s,0}}{(a/a_0)^6}+\frac{\Omega_{\rm rad,0}}{(a/a_0)^4}+\frac{\Omega_{\rm m,0}}{(a/a_0)^3}+\Omega_{\Lambda,0}. \quad (59)$$

The model of late universe considered in this paper is formally equivalent to the ΛCDM model. We could account for a slight deviation to the ΛCDM model (actually the slightest one) by considering a logotropic equation of state [35, 36, 37]. In that case, we obtain a cosmic equation of the form

$$\left(\frac{H}{H_0}\right)^2=\frac{\Omega_{\rm rad,0}}{\frac{\Omega_{\rm rad,0}\rho_0}{\rho_P}+(a/a_0)^4}+\frac{\Omega_{\rm m,0}}{(a/a_0)^3}+\Omega_{\Lambda,0}\left[1+3B\ln(a/a_0)\right], \quad (60)$$

where $B=3.53\times 10^{-3}$ is the so-called dimensionless logotropic temperature [35, 36, 37]. It is given by $B\simeq 1/\ln(\rho_P/\rho_\Lambda)\simeq 1/[123\ln(10)]$. We note that the nonzero value of B is due to quantum mechanics. For $\hbar\to 0$ (i.e. $\rho_P\to +\infty$), we find that $B\to 0$ and we recover the standard ΛCDM model. For $\hbar\neq 0$, the logotropic model is very close to the ΛCDM model (at large scales) up to the present, but will become phantom in about 25 Gyrs and will finally experience a little rip [35, 36, 37]. Owing to this remark and to the remark following Eq. (57), we note that quantum mechanics plays a role both in the early and late universe.

6. Conclusion

We have introduced a model of universe based on a (generalized) polytropic equation of state of the form of Eq. (1). Taking $n=+1$, $k=-4/(3\rho_P)$ and $\alpha=1/3$, we obtain a quadratic equation of state $p/c^2=-4\rho^2/3\rho_P+\rho/3$ that describes the transition between the vacuum energy era (Planck era) and the radiation era. Taking $n=-1$, $k=-\rho_\Lambda$ and $\alpha=0$ we obtain a constant equation of state $p/c^2=-\rho_\Lambda$ that describes the transition between the matter era and the dark energy era (de Sitter era). An interest of our model is that it describes the early universe and the late universe in a symmetric manner. The early universe is described by a polytropic equation of state of index $n=+1$ and the late universe by a polytropic equation of state of index $n=-1$. The first one unifies vacuum energy and radiation and the second one unifies dark matter and dark energy. The mathematical formulae are then strikingly symmetric. In the early universe, this amounts to summing the inverse Planck density and the inverse radiation density ($1/\rho=1/\rho_P+1/\rho_{\rm rad}$). In the late universe, this amounts to summing the matter density and the cosmological density ($\rho=\rho_{\rm m}+\rho_\Lambda$). The Planck density ρ_P in the early universe plays the same role as the cosmological density ρ_Λ in the late universe. They represent fundamental upper and lower density bounds differing by 123 orders of magnitude. They lead to phases of early and late inflation with very different timescales $t_P=1/(G\rho_P)^{1/2}=5.39\times 10^{-44}$s and $t_\Lambda=1/(G\rho_\Lambda)^{1/2}=1.46\times 10^{18}$s. The early universe is governed by quantum mechanics ($\hbar$) and the late universe by general relativity (Λ).

We have also mentioned an alternative model (developed in [23, 24]) where vacumm energy, radiation and dark energy are unified by a single quadratic equation of state $p/c^2=-4\rho^2/3\rho_P+\rho/3-4\rho_\Lambda/3$. This equation of state displays both a phase of early inflation driven by the Planck density ρ_P and a phase of late acceleration driven by the cosmological density ρ_Λ. These two phases are bridged by a phase of deceleration. The pressure is successively negative (vacuum energy), positive (radiation), and negative again (dark energy). In this model, dark matter is treated as an independent species.

We have obtained a cosmic equation (57) that generalizes the ΛCDM model (the ΛCDM model is recovered for $\hbar\to 0$). Our model does not present any singularity and exists "eternally" in the past and in the future, so that it has no origin nor end (we call it aioniotic universe). In particular, the phase of early inflation avoids the Big Bang singularity and replaces it by a sort of second order phase transition where the non zero value of the Planck constant $\hbar$ plays the same role as finite size effects in statistical mechanics [8]. Of course, it is probably incorrect to extrapolate the results in the infinite past since our model is purely classical, or semi-classical, and does not take quantum gravity into account. In our model, the universe starts from $t = -\infty$ but, for $t < 0$, its size is less than the Planck length $l_P = 1.62 \times 10^{-35}$ m. In the Planck era, the classical Einstein equations may be incorrect and should be replaced by a theory of quantum gravity that still has to be constructed.

References

[1] A. Linde, *Particle Physics and Inflationary Cosmology* (Harwood, Chur, Switzerland, 1990)
[2] S. Weinberg, *Gravitation and Cosmology* (John Wiley & Sons, 1972)
[3] E.J. Copeland, M. Sami, S. Tsujikawa, Int. J. Mod. Phys. D **15**, 1753 (2006)
[4] R.R. Caldwell, R. Dave, and P.J. Steinhardt, Phys. Rev. Lett. **80**, 1582 (1998)
[5] A. Kamenshchik, U. Moschella, V. Pasquier, Phys. Lett. B **511**, 265 (2001)
[6] H.B. Sandvik, M. Tegmark, M. Zaldarriaga, I. Waga, Phys. Rev. D **69**, 123524 (2004)
[7] Z.H. Zhu, Astron. Astrophys. **423**, 421 (2004)
[8] P.H. Chavanis, Eur. Phys. J. Plus **129**, 38 (2014)
[9] P.H. Chavanis, Eur. Phys. J. Plus **129**, 222 (2014)
[10] P.H. Chavanis, arXiv:1208.1185
[11] R.R. Caldwell, Phys. Lett. B **545**, 23 (2002)
[12] P.H. Chavanis, AIP Conf. Proc. **1548**, 75 (2013)
[13] S. Tsujikawa, Class. Quantum Grav. **30**, 214003 (2013)
[14] P.H. Chavanis, in preparation
[15] R. R. Caldwell, M. Kamionkowski, N. N. Weinberg, Phys. Rev. Lett. **91**, 071301 (2003)
[16] P.H. Frampton, K.J. Ludwick, R.J. Scherrer, Phys. Rev. D **84**, 063003 (2011)
[17] A. Arvanitaki *et al.*, Phys. Rev. D **81**, 123530 (2010)
[18] A.S. Goldhaber, M.M. Nieto, Rev. Mod. Phys. **82**, 939 (2010)
[19] A. Sen, J. High Energy Phys. **4**, 48 (2002)
[20] G.W. Gibbons, Phys. Lett. B **537**, 1 (2002)
[21] T. Padmanabhan, Phys. Rev. D **66**, 021301(R) (2002)
[22] A. Frolov, L. Kofman, A. Starobinsky, Phys. Lett. B **545**, 8 (2002)
[23] P.H. Chavanis, Journal of Gravity **2013**, 682451 (2013)
[24] P.H. Chavanis, Universe **1**, 357 (2015)
[25] G.A. Monerat *et al.*, Phys. Rev. D **76**, 024017 (2007)
[26] S. Weinberg, Rev. Mod. Phys. **61**, 1 (1989)
[27] T. Padmanabhan, Phys. Rep. **380**, 235 (2003)
[28] Ya. B. Zel'dovich, Soviet Phys. JETP **14**, 1143 (1962)
[29] Ya. B. Zel'dovich, Mon. Not. R. Astron. Soc. **160**, 1 (1972)
[30] P.H. Chavanis, Phys. Rev. D **92**, 103004 (2015)
[31] T. Harko, Mon. Not. R. Astron. Soc. **413**, 3095 (2011)
[32] P.H. Chavanis, Astron. Astrophys. **537**, A127 (2012)
[33] B. Li, T. Rindler-Daller, P.R. Shapiro, Phys. Rev. D **89**, 083536 (2014)
[34] A. Suárez, P.H. Chavanis, Phys. Rev. D **95**, 063515 (2017)
[35] P.H. Chavanis, Eur. Phys. J. Plus **130**, 130 (2015)
[36] P.H. Chavanis, Phys. Lett. B **758**, 59 (2016)
[37] P.H. Chavanis, S. Kumar, JCAP (in press) arXiv:1612.01081

IOP Conf. Series: Journal of Physics: Conf. Series **1030** (2018) 012010 doi:10.1088/1742-6596/1030/1/012010

Cosmologies with scalar fields from higher dimensions: K-essence like fields

J Socorro[1], L Toledo Sesma[1] and L O Pimentel[2]

[1]Departamento de Física, DCeI, Universidad de Guanajuato-Campus León, C.P. 37150, León, Guanajuato, México

[2]Departamento de Física, Universidad Autónoma Metropolitana, Apartado Postal 55-534, C.P. 09340 México, DF, México

E-mail: socorro@fisica.ugto.mx; ltoledo@fisica.ugto.mx; lopr@xanum.uam.mx

Abstract. In this work we construct an effective four-dimensional model by compactifying a ten-dimensional theory of gravity coupled with a real scalar dilaton field on a time-dependent torus a la Kaluza-Klein. The corresponding action in four dimensions is similar to that of K-essence theories. This approach is applied to anisotropic cosmological Bianchi type $\mathrm{VI}_{(h=-1)}$ model for which we study the classical coupling of the anisotropic scale factors with the two real scalar moduli produced by the compactification process, in the epoch to stiff matter with barotropic perfect fluid.

1. Introduction

One of the most important things that we have learned from Planck's results [1] is related to the little anisotropies of the universe. This result have played a tiny role in many theoretical results that describe the dynamics of the universe, if we take the case of the Friedmann-Robertson-Walker (FRW) cosmological model we observe that all the results in this road is developed into the cosmological principle scenario (homogeneous and isotropic universe). Another interesting aspect of cosmology is the inflationary phase that the universe has undergone at early times. In order to explain this process it has been necessary to introduce a scalar field in gravity theory that allows us to explain the accelerated expansion of the universe.

The above issues have suggested to consider the presence of higher-dimensional degrees of freedom on the cosmology derived from four-dimensional effective theories. In the higher-dimensional theories an effective action with a graviton and a massless scalar field is considered, the dilaton is one component driving the evolution of the universe [2, 3].

In the present work we shall consider an alternative procedure about the role played by the moduli. In particular we shall not consider the presence of fluxes, as in string theory, in order to obtain a moduli-dependent scalar potential in the effective theory. Rather, we are going to promote some of the moduli to time-dependence by considering the particular case of a ten-dimensional gravity coupled to a time-dependent dilaton compactified on a six-dimensional torus with a time-dependent Kähler modulus. With the propose to track down the role play by such fields, we are going to ignore the dynamics of the complex structure field (for instance, by assuming that it is already stabilized by the presence of a string field in higher scales).

IOP Conf. Series: Journal of Physics: Conf. Series **1030** (2018) 012010 doi:10.1088/1742-6596/1030/1/012010

2. Effective model

We start from a ten-dimensional gravity theory coupled with a dilaton. In the string frame, the effective action depends on two space-time-dependent scalar fields: the dilaton $\Phi(x^\mu)$ and the Kähler modulus $\sigma(x^\mu)$. For simplicity, in this work we shall assume that these fields only depend on time. The higher dimensional (effective) theory is therefore given by

$$S = \frac{1}{2\kappa_{10}^2}\int d^{10}X\sqrt{-\hat{G}}\, e^{-2\Phi}\left[\hat{\mathcal{R}}^{(10)} + 4\,\hat{G}^{MN}\nabla_M\,\Phi\,\nabla_N\,\Phi\right] + \int d^{10}X\,\sqrt{-\hat{G}}\,\hat{\mathcal{L}}_m \qquad (1)$$

$\hat{q}$ refer to the string frame, where q are any quantity that appear in the any equation in this work, while the ten-dimensional metric is described by

$$ds^2 = \hat{G}_{MN}\,dX^M\,dX^N = \hat{g}_{\mu\nu}\,dx^\mu dx^\nu + h_{mn}\,dy^m dy^n, \qquad (2)$$

the six-dimensional internal space has the form of a torus with a metric given by

$$h_{mn} = e^{-2\sigma(t)}\,\delta_{mn}, \qquad (3)$$

with σ a real time dependent parameter.

By reducing, a la kaluza-Klein, the higher dimensional action (1) to four dimensions and rewriting it in the Einstein frame we have

$$S_4 = \frac{1}{2\kappa_4^2}\int d^4x\sqrt{-g}\left(R - 2\,g^{\mu\nu}\,\nabla_\mu\phi\nabla_\nu\phi - 96\,g^{\mu\nu}\,\nabla_\mu\sigma\nabla_\nu\sigma - 36\,g^{\mu\nu}\,\nabla_\mu\phi\nabla_\nu\sigma + \mathcal{L}_m^{(4)}\right), \qquad (4)$$

which is similar to the K-essence field action, in the sense that the scalar field potential does not appear in the action. Many work has been done in order to obtain classical solutions using Hamilton formalism or Hamilton-Jacobi equations [4]. Many relations in the compactification mode can be seen in references [2, 3].

2.1. Effective Einstein equations in four dimensions

The equation of motions associated to the reduced action in four dimensions (4) can be obtained by taking variations with respect to each of the fields. Then we have that the Einstein equations and the Klein-Gordon like equations (EKG) for to the scalar fields (ϕ, σ) are given by

$$\begin{aligned} R_{\alpha\beta} - \frac{1}{2}\,g_{\alpha\beta}\,R &= 2\left(\nabla_\alpha\phi\,\nabla_\beta\phi - \frac{1}{2}\,g_{\alpha\beta}\nabla^\gamma\,\phi\nabla_\gamma\phi\right) + 96\,\left(\nabla_\alpha\sigma\,\nabla_\beta\sigma - \frac{1}{2}\,g_{\alpha\beta}\,\nabla^\gamma\sigma\,\nabla_\gamma\sigma\right) \\ &\quad + 36\,\left(\nabla_\alpha\phi\,\nabla_\beta\sigma - \frac{1}{2}\,g_{\alpha\beta}\,\nabla^\gamma\phi\,\nabla_\gamma\sigma\right) + 8\pi G\,T_{\alpha\beta}, \qquad (5) \\ \Box\phi &= 0, \qquad (6) \\ \Box\sigma &= 0, \qquad (7) \end{aligned}$$

where $\Box$ is the d'Alembertian operator in four dimensions and $T_{\alpha\beta}$ is the effective energy-momenta tensor corresponding to barotropic perfect fluid in four dimensions.

In this work we consider an inisotropic universe and use the metric for the Bianchi type $VI_{h=-1}$ model with the following line element

$$ds^2 = -N^2 dt^2 + A^2(t)\,dx^2 + B^2(t)e^{-2x}\,dy^2 + C^2(t)e^{2x}\,dz^2, \qquad (8)$$

where N(t) is the lapse function, the functions A(t), B(t) and C(t) are scale factors in the directions (x, y, z), respectively.

Introducing the metric (8), into (5), the corresponding Einstein-Klein-Gordon equations are

$$G^0_0: \quad 2\frac{A'}{A}\frac{C'}{C} + \left(\frac{C'}{C}\right)^2 - \phi'^2 - 18\phi'\sigma' - 48\sigma'^2 - 8\pi G\rho - \frac{1}{A^2} = 0, \tag{9}$$

$$G^0_1 = -G^1_0: \quad \frac{B'}{B} - \frac{C'}{C} = 0, \tag{10}$$

$$G^1_1: \quad 2\frac{C''}{C} + \left(\frac{C'}{C}\right)^2 + \phi'^2 + 18\phi'\sigma' + 48\sigma'^2 + 8\pi GP + \frac{1}{A^2} = 0, \tag{11}$$

$$G^2_2: \quad \frac{A''}{A} + \frac{A'}{A}\frac{C'}{C} + \frac{C''}{C} + \phi'^2 + 18\phi'\sigma' + 48\sigma'^2 + 8\pi GP - \frac{1}{A^2} = 0, \tag{12}$$

$$G^3_3: \quad \frac{A''}{A} + \frac{A'}{A}\frac{B'}{B} + \frac{B''}{B} + \phi'^2 + 18\phi'\sigma' + 48\sigma'^2 + 8\pi GP - \frac{1}{A^2} = 0, \tag{13}$$

$$\Box\phi = 0: \quad \left(\frac{A'}{A} + 2\frac{C'}{C}\right)\phi' + \phi'' = 0, \tag{14}$$

$$\Box\sigma = 0: \quad \left(\frac{A'}{A} + 2\frac{C'}{C}\right)\sigma' + \sigma'' = 0, \tag{15}$$

where $\prime = \frac{d}{d\tau} = \frac{d}{Ndt}$.

From the last expressions, we can see that it is easy to solve equation (10) whose solution for simplicity is $B(t) = C(t)$, with this result, the metric present a cylindrical symmetry in the plane (y,z), that we rename a hidden symmetry for this cosmological model. Also, the solution for the fields (ϕ, σ) are obtained from equations (14), and (15)

$$\phi(\tau) = \phi_0 \int \frac{d\tau}{AC^2} + \phi_1, \qquad \sigma(\tau) = \sigma_0 \int \frac{d\tau}{AC^2} + \sigma_1. \tag{16}$$

By taking the relation between the proper and cosmic time ($d\tau = N(t)dt$) we see that the lapse function $N(t)$ can be choose as $N \sim AC^2$. This tell us that the lapse function plays the role of a gauge transformation. Then with that choice for N it is possible to obtain the simplest solution in the cosmic time t, for the scalar fields as

$$\phi(\tau) = \phi_0 t + \phi_1, \qquad \sigma(\tau) = \sigma_0 t + \sigma_1, \tag{17}$$

where $(\phi_0, \phi_1, \sigma_0, \sigma_1)$ are integration constants. With this behavior in the scalar fields, the radii in the internal metric (3) goes to zero in dynamical manner. In this way, in the observed dimensions, the extra dimensions can not be observed today.

So far, we have found from the fields equations (10) that the radii B and C are equal for the Bianchi $VI_{h=-1}$, and the moduli fields (ϕ, σ) have linear behavior in time, as can be seen in the expressions (17). These solutions were possible obtained by taking the gauge $N \sim AC^2$.

To obtain the solutions for the radii A and C, we will use an alternative method, namely the Hamilton approach, with the idea that these solutions *must be solutions to Einstein-Klein-Gordon equations.*

3. Classical Lagrangian and Hamiltonian

With the idea to reach the quantum scheme of our model, we shall develop the classical Lagrangian and Hamiltonian analysis. Then, we include the Lagrangian density for the matter content (barotropic perfect fluid) of the universe

$$\mathcal{L}_{VI} = 2\frac{A\dot{C}^2}{N} + 4\frac{C\dot{A}\dot{C}}{N} + 2\frac{AC^2\dot{\phi}^2}{N} + 36\frac{AC^2\dot{\phi}\dot{\sigma}}{N} + 96\frac{AC^2\dot{\sigma}^2}{N} + 2\frac{C^2}{A}N + 16\pi GAC^2\rho N, \tag{18}$$

in the last expression we have employed the result that the radii B and C are equal.

Using the standard definition of the momenta $\Pi_{\mathrm{q}^\mu} = \frac{\partial \mathcal{L}}{\partial \dot{\mathrm{q}}^\mu}$, where q^μ are the coordinate fields $q^\mu = (A, C, \phi, \sigma)$ we obtain the momenta associated to each field

$$\begin{aligned}
\Pi_{\mathrm{A}} &= \frac{4\mathrm{C}\dot{\mathrm{C}}}{\mathrm{N}}, & \dot{\mathrm{C}} &= \frac{\mathrm{N}\Pi_{\mathrm{A}}}{4\mathrm{C}}, \\
\Pi_C &= 4\frac{A\dot{C}}{N} + 4\frac{C\dot{A}}{N}, & \dot{A} &= \frac{N}{4AC^2}\left[C\Pi_C - A\Pi_A\right], \\
\Pi_\phi &= \frac{AC^2}{N}\left[36\dot{\sigma} + 4\dot{\phi}\right], & \dot{\phi} &= \frac{N}{44AC^2}\left[3\Pi_\sigma - 16\Pi_\phi\right], \\
\Pi_\sigma &= \frac{AC^2}{N}\left[192\dot{\sigma} + 36\dot{\phi}\right], & \dot{\sigma} &= \frac{N}{132AC^2}\left[9\Pi_\phi - \Pi_\sigma\right],
\end{aligned}$$

and introducing them into the Lagrangian density, we obtain the canonical Lagrangian as $\mathcal{L}_{canonical} = \Pi_{q^\mu}\dot{q}^\mu - N\mathcal{H}$. When we perform the variation of this canonical Lagrangian with respect to N, $\frac{\delta \mathcal{L}_{canonical}}{\delta N} = 0$, we obtain the constraint $\mathcal{H} = 0$.

Therefore we obtain the Hamiltonian density for this model as follows

$$\mathcal{H}_{\mathrm{VI}} = -\mathrm{A}^2\Pi_{\mathrm{A}}^2 + 2\mathrm{A}\Pi_{\mathrm{A}}\,\mathrm{C}\Pi_{\mathrm{C}} - \frac{16}{11}\Pi_\phi^2 + \frac{6}{11}\Pi_\phi\,\Pi_\sigma - \frac{1}{33}\Pi_\sigma^2 - 128\,\pi\,\mathrm{G_N}\,\rho_\gamma\left(\mathrm{AC}^2\right)^{1-\gamma} - 16\mathrm{C}^4, \quad (19)$$

where we used the gauge $N = 8AC^2$. The last Hamiltonian can be rewritten in a simpler way by taking the transformations related with the generalized coordinates and momenta as $\mathrm{Q} = \mathrm{e}^{\mathrm{q}}$, and $\Pi_{\mathrm{Q}} = \frac{\partial \mathrm{S}}{\partial \mathrm{Q}}$, where S is the superpotential function in the Hamilton-Jacobi formalism. The momenta can be transformed as follow $\Pi_{\mathrm{A}} \to \Pi_{\mathrm{A}} = \mathrm{e}^{-\mathrm{a}}\mathrm{P_a}$, where $\mathrm{P_a} = \frac{\partial \mathrm{S}}{\partial \mathrm{a}}$ and we write in a similar way the other momenta in the field (ϕ, σ), that is $\Pi_\phi = \mathrm{P}_\phi$, $\Pi_\sigma = \mathrm{P}_\sigma$.

Now, we develop our analysis for the case when the matter content is taken to be a stiff barotropic fluid, $\gamma = 1$, and the Hamiltonian density is

$$\mathcal{H} = -\mathrm{P_a^2} + 2\mathrm{P_a}\mathrm{P_c} - \frac{16}{11}\mathrm{P}_\phi^2 + \frac{6}{11}\mathrm{P}_\phi\,\mathrm{P}_\sigma - \frac{1}{33}\mathrm{P}_\sigma^2 - 16\mathrm{e}^{4\mathrm{c}} - 128\,\pi\,\mathrm{G_N}\,\rho_1, \quad (20)$$

the corresponding Hamilton equations for the momenta and coordinates are

$$\begin{aligned}
\dot{\mathrm{a}} &= -2\mathrm{P_a} + 2\mathrm{P_c}, && && (21)\\
\dot{\mathrm{c}} &= 2\mathrm{P_a}, && && (22)\\
\dot{\phi} &= -\frac{32}{11}\mathrm{P}_\phi + \frac{6}{11}\mathrm{P}_\sigma, && && (23)\\
\dot{\sigma} &= \frac{6}{11}\mathrm{P}_\phi - \frac{2}{33}\mathrm{P}_\sigma, && && (24)\\
\dot{\mathrm{P}}_{\mathrm{a}} &= 0, & \Rightarrow\quad & \mathrm{P_a} = \mathrm{p_a} = \text{constant}, && (25)\\
\dot{\mathrm{P}}_{\mathrm{c}} &= 64\mathrm{e}^{4\mathrm{c}}, && && (26)\\
\dot{\mathrm{P}}_\phi &= 0, & \Rightarrow\quad & \mathrm{P}_\phi = \mathrm{p}_\phi = \text{constant}, && (27)\\
\dot{\mathrm{P}}_\sigma &= 0, & \Rightarrow\quad & \mathrm{P}_\sigma = \mathrm{p}_\sigma = \text{constant}. && (28)
\end{aligned}$$

The last system of differential equations is solved by

$$\mathrm{c(t)} = 2\mathrm{p_a}\,\mathrm{t} + \mathrm{c_o}, \qquad \to \mathrm{C(t)} = \mathrm{C_0}\mathrm{e}^{2\mathrm{p_a}\mathrm{t}}$$

.

$$\mathrm{a(t)} = \mathrm{a_0} + \frac{\mathrm{b_0}}{4\mathrm{p_a}}\mathrm{t} + \frac{\mathrm{p_0}}{32\mathrm{p_a^2}}\mathrm{e}^{8\mathrm{p_a}\mathrm{t}}, \qquad \mathrm{A(t)} = \mathrm{A_0}\mathrm{e}^{\frac{\mathrm{b_0}}{4\mathrm{p_a}}\mathrm{t}}\,\exp\left[2\frac{\mathrm{C_0^4}}{\mathrm{p_a^2}}\mathrm{e}^{8\mathrm{p_a}\mathrm{t}}\right].$$

$$\phi(\tau) = \phi_0 \mathrm{t} + \phi_1, \qquad \sigma(\tau) = \sigma_0 \mathrm{t} + \sigma_1,$$

where (ϕ_1, σ_1) are integration constants, whereas $\phi_0 = -\frac{32}{11}\mathrm{p}_\phi + \frac{6}{11}\mathrm{p}_\sigma$ and $\sigma_0 = \frac{6}{11}\mathrm{p}_\phi - \frac{2}{33}\mathrm{p}_\sigma$.

These solutions were introduced into the Einstein field equations, and they are satisfied, with $b_0 = \phi_0^2 + 18\phi_0\sigma_0 + 512\pi G\rho_1 - 4p_a^2 + 48\sigma_0^2$, and $P_0 = 64C_0^4$.

4. conclusions

In this work we have developed the anisotropic Bianchi $\mathrm{VI}_{\mathrm{h}=-1}$ model from a higher-dimensional theory of gravity, our analysis cover the classical aspects and prepared us to tackle the quantum version of this work. In the frame of classical analysis we use the tools of the analytical mechanics for solve these, in particular way, the Hamilton equations for solve the corresponding set of equations, these solutions were checked over the original Einstein-Klein-Gordon equations. We can see that in the reduced theory from a higher-dimensional theory to four-dimensions, two radii of the Bianchi $VI_{h=-1}$ type are equals (B(t) = C(t)) presenting a hidden symmetry, which says that the volume function of this cosmological model goes as a jet in the x direction, whereas in the (y,z) plane goes as a circle (cylindrical symmetry in this plane). This hidden symmetry allows us to simplify the analysis in the classical context, and therefore also in the quantum version. When we introduce this classical hidden symmetry in the quantum version for this model in the Bohm's formalism, the corresponding quantum solution is simplify [5], in sense that the changes introduced 21 years ago for the Bianchi's Class A cosmological model for this model in the matrix m^{ij} representation in the classification scheme of Ellis and MacCallum [6], is eliminated. In the time behavior of the scalar fields does that the radii in the internal metric (3) goes to zero in dynamical manner, thus, the extra dimensions can not be observed today.

Acknowledgments

This work was partially supported by CONACYT 167335, 179881 grants. PROMEP grants UGTO-CA-3 and UAM-I-43. This work is part of the collaboration within the Instituto Avanzado de Cosmología and Red PROMEP: Gravitation and Mathematical Physics under project *Quantum aspects of gravity in cosmological models, phenomenology and geometry of space-time*. Many calculations where done by Symbolic Program REDUCE 3.8.

References

[1] Ade, P. A. R. et al, Planck Collaboration, Astron. Astrophys. **594**, A18 (2016),[arXiv:1502.01593], *Planck 2015 results - XVIII. Background geometry and topology of the Universe.*

[2] Socorro, J. and Toledo Sesma, L; Eur. Phys. J. Plus **131**, 71 (2016), *Time-dependent toroidal compactification proposals and the Bianchi type II model: classical and quantum solutions*, [arXiv:1507.03171].

[3] Sesma, L. Toledo and Socorro, J. and Loaiza, O., Adv. High Energy Phys. 6705021 (2016), *Time-dependent toroidal compactification proposals and the Bianchi type I model: Classical and quantum solutions*, [arXiv:1501.02779].

[4] Socorro, J. and Pimentel Luis O., and Espinoza García, Abraham, Advances in High Energy Phys. 805164 (11p) (2014), *Classical Bianchi type I cosmology in K-essence theory*

[5] Obregón, O. and Socorro, J; Int. J. Theor. Phys. **35**, 1381 (1996), $\Psi = \mathrm{W}e^{\pm\phi}$ *quantum cosmological solutions for Class A Bianchi models.*

[6] Ryan, M.P., and Shepley, L.C.;*Homogeneous Relativistic Cosmologies*, Princeton University Press, Princeton, New Jersey (1975),

IOP Conf. Series: Journal of Physics: Conf. Series **1030** (2018) 012011 doi:10.1088/1742-6596/1030/1/012011

Stealth field in FLRW spacetime with non minimal derivative coupling

Gilberto Aguilar, Abigail Alvarez and Cuauhtemoc Campuzano.

Facultad de Física, Universidad Veracruzana, México

E-mail: zs15028854@estudiantes.uv.mx, aalvarez@uv.mx and ccampuzano@uv.mx

Abstract. In this work, we show a solution for the stealth scalar field arising from a non-minimal derivative coupling in a Friedmann-Lemaître-Robertson-Walker (FLRW) spacetime coupled to the simplest case of a perfect fluid, namely, dust.

IOP Conf. Series: Journal of Physics: Conf. Series **1030** (2018) 012011 doi:10.1088/1742-6596/1030/1/012011

1. Introduction

Recently in modified theories of gravitation, particularly in special cases of Hordenski theory it has been studied that these have implications on cosmological models in order to provide late mechanisms of self tuning [1]. As it was reported in [1] the theory rised by the Lagrangians L_{jhon} and L_{paul} provide a self tuning mechanism, being these the strongest terms of the so-called four Fabs. Within the context of the modified theories and the Hordenski one, there are interesting solutions for scalar fields that emerge from theories with non minimal coupling between the scalar fields and gravity, named *stealths*. The main feature of the stealth is that they do not backreact on gravitational fields, so its existence leave intact the evolution of underlying spacetime. As is well mentioned in [2], stealths may be perceived as a mathematical curiosity, however, their behavior could alleviate the cosmological constant problem: basically the predictions about the energy density of the quantum vacuum is approximately 120 orders of magnitude larger than the measured cosmological energy density. Also in the same reference, it was analyzed the stability of solutions finding three stables cases against five unstable ones and showing the sensibility of the integration constants and the coupling parameter in the analyzed cases.

The stealth fields are being studied in different contexts. For instance, in vector-tensor theories [3, 4], three dimensional black-holes [5, 6], Galileons [7]. The existence of these configurations have been found in [6] for the well known $(2+1)$-dimensional BTZ static black hole and later on in $(3+1)$-dimensions, for the Minkowski flat spacetime [8], in (A)dS spacetime [9]; in the cosmology context [10, 11, 12, 13], and on a charged dilatonic $(1+1)$-D black hole [14]. In a similar spirit, in n dimensions, for higher-order gravity theories [15, 16], in the Lovelock gravity context [17] as well as in the non-relativistic version of the gauge/gravity correspondence in Lifshitz spacetime [18].

In this work, we study the term L_{john} non minimally coupled to the scalar field in order to find solutions for stealth fields in a Friedmann-Lemaître-Robertson-Walker spacetime in presence of sources, choosing as a source the simplest case of a perfect fluid, namely, dust. That is consistent with the scenario described in [19] about the Universe modeled as a fluid and its evolution going over different stages. From recent observations, it is known that ordinary matter accounts for $4\% \pm 1$ of the total mass/energy content, with dark matter and dark energy accounting 96%; and its expansion is accelerating and not decelerating. So in a good approximation, the stress energy-momentum tensor in the present Universe is described by the dust energy-momentum tensor.

2. Field equations and stealth configurations

As is well known, the cosmological principle suggest a homogeneous and isotropic Universe and its evolution is given by the cosmological standard model. A spacetime which agrees with this principle is the line element Friedmann-Lemaître-Robertson-

Walker;

$$ds^2 = -dt^2 + a^2(t)\left[\frac{dr^2}{1-kr^2} + r^2(d\theta^2 + sen^2\theta d\varphi^2)\right]. \quad (1)$$

where $a(t)$ is the scale factor, and the constant k determines the topology of the Universe, for $k = 0$ flat, $k = -1$ open and $k = 1$ closed Universe. In the dust case $p = 0$, the energy density evolves as $\rho = \rho_0 a^{-3}$ and its scale factor has the functional form $a(t) = a_0 t^{2/3}$.

Stealth configurations have been found when one considers some coupling between a scalar field and gravity, which is described by the action

$$S[g_{\mu\nu}, \phi] = \int d^4x\sqrt{-g}\left(\frac{1}{2\kappa}R + \frac{1}{2}(\eta G_{\mu\nu} - g_{\mu\nu})\nabla^\mu\phi\nabla^\nu\phi - V(\phi)\right) \quad (2)$$

where R is the Ricci scalar, $G_{\mu\nu}$ is the Einstein tensor, $\phi(t)$ is a scalar field, $V(\phi)$ is the scalar field's potential, and η is the coupling parameter, from this action the term L_{john} is easily recognized.

After performing the variation with respect to metric, the field equations can be rewritten as,

$$G_{\mu\nu} - \kappa T^m_{\mu\nu} = \kappa T^s_{\mu\nu}, \quad (3)$$

where $T^m_{\mu\nu}$ and $T^s_{\mu\nu}$ are the energy-momentum tensor of matter and stealth, respectively.

Explicitly, the stealth stress-energy tensor for the non minimal derivative coupling

$$\begin{aligned} T^s_{\mu\nu} &= \nabla_\mu\phi\nabla_\nu\phi - \left[\frac{1}{2}\nabla_\lambda\phi\nabla^\lambda\phi + V(\phi)\right]g_{\mu\nu} + \eta\left\{\frac{1}{2}\nabla_\mu\phi\nabla_\nu\phi R\right. \\ &- 2\nabla_\lambda\phi\nabla_{(\mu}\phi R^\lambda_{\nu)} - \nabla^\lambda\phi\nabla^\rho\phi R_{\mu\lambda\nu\rho} - (\nabla_\mu\nabla^\lambda\phi)(\nabla_\nu\nabla_\lambda\phi) \\ &+ (\nabla_\mu\nabla_\nu\phi)\Box\phi + \frac{1}{2}G_{\mu\nu}(\nabla\phi)^2 - g_{\mu\nu}\left[-\frac{1}{2}(\nabla^\lambda\nabla^\rho\phi)(\nabla_\lambda\nabla_\rho\phi)\right. \\ &\left.\left.+ \frac{1}{2}(\Box\phi)^2 - \nabla_\lambda\phi\nabla_\rho\phi R^{\lambda\rho}\right]\right\} = 0, \end{aligned} \quad (4)$$

the stealth solutions are found imposing the vanishing of the above expression. By inserting (1) into (4), one finds that the system of equations which leaves (4) are just two linearly independent components, namely

$$T_t{}^t = \frac{1}{2}(9\eta H^2 - 1)\left(\frac{d\phi}{dt}\right)^2 - V(\phi), \quad (5)$$

$$T_r{}^r = \frac{1}{2}[\eta(2H' + 3H^2) + 1]\left(\frac{d\phi}{dt}\right)^2 + \eta H\frac{d}{dt}\left(\frac{d\phi}{dt}\right)^2 - V(\phi), \quad (6)$$

here $H = \frac{a'}{a}$ is the Hubble parameter and H' its temporal derivate.

The problem now is to solve the system of equations given by (5) and (6) for ϕ and $V(\phi)$, so combining both components we have the equation for the scalar field

$$\frac{1}{\eta H}(T_r{}^r - T_t{}^t) = \frac{d}{dt}\left(\frac{d\phi}{dt}\right)^2 + \left(\frac{\eta H' - 3\eta H^2 + 1}{\eta H}\right)\left(\frac{d\phi}{dt}\right)^2 = 0, \quad (7)$$

which integrated for $(\frac{d\phi}{dt})^2$ gives the solution

$$\left(\frac{d\phi(t)}{dt}\right)^2 = \int \exp\left(-\frac{\eta H' - 3\eta H^2 + 1}{\eta H}\right) dt = \phi_0 t^3 \exp\left(-\frac{3t^2}{4\eta}\right) \tag{8}$$

where $\phi_0 > 0$ is an integration constant. Integrating the above equation we get;

$$\phi(t) = -\frac{2}{9}\eta(3t^2 + 4\eta)\exp(-\frac{3}{4\eta}t^2) + \frac{8}{9}\eta^2. \tag{9}$$

Now, in order to find the self-interacting potential for the scalar field, we consider the combination of components

$$T_r{}^r + T_t{}^t = \eta H \frac{d}{dt}\left(\frac{d\phi}{dt}\right)^2 + \eta\left(H' + 6H^2\right)\left(\frac{d\phi}{dt}\right)^2 - 2V(\phi) = 0, \tag{10}$$

from it, we isolate the potential $V(\phi)$. After the substitution of (8) into (10) one obtains

$$V(\phi) = \frac{1}{2}(9\eta H^2 - 1)\left(\frac{d\phi}{dt}\right)^2 = \frac{1}{2}\left(\frac{4\eta}{t^2} - 1\right)\phi_0 t^3 \exp\left(-\frac{3t^2}{4\eta}\right), \tag{11}$$

and it is worth mentioning that this coincides with the component $T_t{}^t$.

Now in this case it is possible to write the self-interaction potential just in terms of ϕ, that is possible by inverting the Eq.(9), i.e., taking the inverse function in order to get $t = t(\phi)$:

$$t = -\frac{2}{9}\sqrt{-3\eta}\sqrt{LambertW(\hat{\phi}) + 1}, \tag{12}$$

here $\hat{\phi} = (\frac{9}{8\eta^2}\phi - 1)e^{-1}$. Hence, the selft-interaction potential finally is written as

$$V(\phi) = \frac{\phi_0\sqrt{-3\eta}}{18\eta}\frac{(LambertW(\hat{\phi}) + 4)\sqrt{LambertW(\hat{\phi}) + 1}}{LambertW(\hat{\phi})}(8\eta^2 - 9\phi), \tag{13}$$

there are two points to discuss from the potential (13), first, in order to avoid complex values for the potential, η must be in the range $\eta \in (-\infty, 0)$, at second point is the LambertW‡ function; the principal branch is analytic at zero and it has real-values for the range $\phi \in [-e^{-1}, \infty)$, that is our case, so the potential is real and well behaved in the range $V(\phi) \in [0, \infty)$.

In the figure (1) we show the graph of $V(\phi)$ vs. ϕ, in order to show the behavior of the potential, actually, it is not so sensitive to the values of parameters, so that for any values of these, the graph will be similar.

‡ The LambertW function, satisfies $LambertW(x)e^{LambertW(x)=x}$, as the equation $ye^y = x$ has an infinite number of solutions y for each (non-zero) value of x, so has an infinity number of branches, but just one analytic at 0.

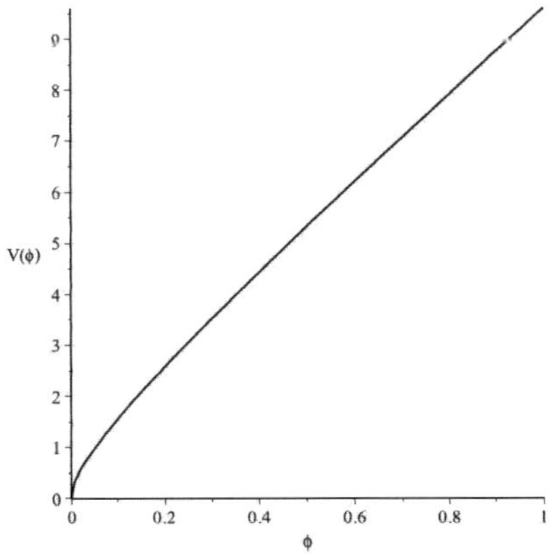

Figure 1. Self-interaction potential $V(\phi)$, for $\phi_0 = 1$ and $\eta = -1/4$.

3. Conclusions

The aim of the present work was to show the existence of solutions in theories with non-minimal derivative coupling– motivated by the self-tuning of this coupling and the behavior of stealth fields–, in order to study its possible cosmological implications. In particular, we studied the simplest case of a perfect fluid (dust) in an homogeneous and isotropic cosmology described by the FLRW spacetime. All the development shown the decoupling of the field equations to get the explicit integral of stealth field, and its self-interaction potential.

On the other hand, in the looking for stealth solutions, one can follow different philosophies, fixing the self-interaction potential, as in [12], and decoupling the equations to find the field, or as in [19], fixing the cosmology to find the field an its potential as is our case. When the cosmology was appointed, the most often is founding dissipative potentials and in order to get a potential just in terms of the scalar field one use the inverse function of the implicit variable to obtain it in terms of the scalar field.

Although the approach used in the present work helped us to found stealth solutions, it is not direct to obtain the self-tunning mechanism. Instead, there is an approach in which the stealth solutions and self-tuning mechanism are obtained directly [20, 21]. Unfortunately, the recent observations discard this particular coupling term [22], however, the study of all the approaches emerging from it could be interesting.

Acknowledgments

The authors acknowledge partial support by CONACyT Grant CB-2012-177519-F. This work was partially supported by SNI (México). AA acknowledges partial support by CONACyT Grant Estancias Posdoctorales Vinculadas al Fortalecimiento de Calidad del Posgrado Nacional 2016.

References

[1] Charmousis, C., Edmund J. Copeland,E. J., Padilla, A., Paul M. Saffin, P. M.: General second order scalar-tensor theory, self tuning, and the Fab Four. Phys. Rev. Lett. **108** 051101 (2012). arxiv: 1106.2000 [hep-th]

[2] Faraoni, V., Zambrano-Moreno, A. F.: Are stealth scalar fields stable? Phys. Rev. D **81** 124050 (2010). arxiv:1006.1936 [gr-qc]

[3] Chagoya, J., Tasinato, G.: Stealth configurations in vector-tensor theories of gravity. JCAP **1801** no. 1 046 (2018). arxiv:1707.07951 [hep-th]

[4] Chagoya, , Niz, G. and Tasinato, G.: Black Holes and Abelian symmetry breaking. Class.Quant.Grav. **33** no.17, 175007 (2016). arxiv.1602.08697

[5] Cárdenas, M., Fuenteaba, O. and Martínez, C.: Three-dimensiona black holes with conformal y coupled scalar field. Phys.Rev. D **90** no.12, 124072 (2014). arxiv:1408.1401 [hep-th]

[6] Ayón-Beato, E., Martínez, C., Zanelli, J.: Stealth scalar field overflying a $(2+1)$ black hole. Gen. Rel. Grav. **38**, 145 (2006). arXiv: hep-th/0403228

[7] Babichev, E., Charmousis, C.: Dressing a black hole with a time-dependent Galileon. JHEP 1408, 106 (2014). arXiv:1312.3204 [gr-qc]; Babichev, E., Charmousis, C., Hassaine, M.: Charged Galileon black holes. JCAP 1505, 031 (2015). arXiv:1503.02545 [gr-qc]

[8] Ayón-Beato, E., Martínez, C., Troncoso, R., Zanelli, J.: Gravitational Cheshire effect: Nonminimally coupled scalar fields may not curve spacetime. Phys. Rev. D **71**, 104037 (2005). arXiv: hep-th/0505086

[9] Ayón-Beato, E., Martínez, C., Troncoso, R., Zanelli, J.: Stealths on AdS. Private communication.

[10] N. Banerjee, R. K. Jain and D. P. Jatkar, Gen. Rel. Grav. **40**, 93 (2008) doi:10.1007/s10714-007-0516-1 [hep-th/0610109].

[11] Maeda, H., Maeda, K. I.: Creation of the universe with a stealth scalar field. Phys. Rev. D **86**, 124045 (2012). arXiv:1208.5777 [gr-qc]

[12] Ayón-Beato, E., García, A. A., Ramírez-Baca, P. I., Terrero-Escalante, C. A.: Conformal stealth for any standard cosmology. Phys. Rev. D **88**, 063523 (2013). arXiv:1307.6534 [gr-qc]

[13] Campuzano, C., Cárdenas, V. H. and Herrera, R.:Mimicking the LCDM model with Stealths, Eur.Phys.J. C **76** no.12, 698 (2016). arXiv:1611.08433 [gr-qc].

[14] A. Alvarez, C. Campuzano, M. Cruz, E. Rojas and J. Saavedra, Gen. Rel. Grav. **48** (2016) no.12, 165 doi:10.1007/s10714-016-2158-7 [arXiv:1611.03022 [gr-qc]].

[15] Bravo-Gaete, M., Hassaine, M.: Topological black holes for Einstein-Gauss-Bonnet gravity with a nonminimal scalar field. Phys. Rev. D **88**, 104011 (2013). arXiv:1308.3076 [hep-th]

[16] Hassaine, M.: Rotating AdS black hole stealth solution in $D = 3$ dimensions. Phys. Rev. D **89**, 044009 (2014). arXiv:1311.4623 [hep-th]

[17] Bravo-Gaete, M., Hassaine, M.: Planar AdS black holes in Lovelock gravity with a nonminimal scalar field. JHEP **1311**, 177 (2013). arXiv:1309.3338 [hep-th]

[18] Ayón-Beato, E., Hassaine, M., Juárez-Aubry, M. M.: Stealths on Anisotropic Holographic Backgrounds. arXiv:1506.03545 [gr-qc]

[19] Ayón-Beato, E., Ramírez-Baca, P. I., Terrero-Escalante, C. A.: Conformal stealth with no conformal coupling. Phys. Rev. D **97**, 043505 (2018). arXiv:1512.09375 [gr-qc]

[20] Aguilar, G. Tesis de Maestría: Algunas consecuencias cosmológicas de los campos stealth, Universidad Veracruzana (2018).

[21] Alvarez, A, Campuzano, C, Cárdenas, V., Rojas, E. Some stealth's cosmological consequences. *in preparation*; Aguilar, G, Campuzano, C., Cárdenas, V.: stealth's self-tunning mechanism. *in preparation*

[22] Ezquiaga, J. M., Zumalacárregui, M.: Dark Energy after GW170817: dead ends and the road ahead, Phys. Rev. Lett, **119** no. 25, 251304 (2017); Creminelli, P., Vernizzi, F. Phys. Rev. Lett. **119** no. 25, 251302 (2017).

IOP Conf. Series: Journal of Physics: Conf. Series **1030** (2018) 012012 doi:10.1088/1742-6596/1030/1/012012

Hidden SUSY features, and Dawson's function and the wrong-sign Hermite functions

Marco A. Reyes and Rafael Arcos-Olalla

Departamento de Física, Universidad de Guanajuato, México

E-mail: marco@fisica.ugto.mx

Abstract. We propose that some problems in mathematical physics show non-typical supersymmetric (SUSY) features which may still need to be thoroughly explored. In particular, we first show that the Pöschl-Teller problem posses some inherent ladder peculiarities which are not considered in the usual SUSY approach. Then we show that Dawson's function and its derivatives show a SUSY type relation with the wrong-sign Hermite equation, found in a general factorization of the simple harmonic oscillator, which includes Mielnik's SUSY partners as particular cases.

1. Introduction

One of the basic problems when trying to find analytical solutions to dynamic problems, is the fact that only a very small number of exact solutions to second order ordinary differential equations are known, the number being smaller if one wishes to work with non-linear problems. In the case of one-dimensional quantum mechanics (QM), only about a dozen of exact solutions are known, and therefore the appearance of supersymmetric (SUSY) quantum mechanics (QM) techniques in the 1980's [1] was immediately embraced to find at least new problems with known old spectra. Work in this direction developed so fast that nowadays it seems that everything has been solved, leaving SUSY applications to cosmological problems, where new solutions are still being found.[2] However, as is usual with fast growing fields in science, some details in the subject were not thoroughly explored, and one finds from time to time new characteristics forgotten or hidden that still need to be solved out. Here we show how some SUSY aspects were forgotten in the settled work, and that some surprises can still be found that may deserve our attention.

We begin our article with a discussion of the Pöschl-Teller QM problem, to motivate the argument that some forgotten information may still need consideration. Afterwards, we show how new SUSY type factorizations lead to surprises, as can be seen in the case of the wrong-sign Hermite equation and the differential equation for Dawson's function's derivatives.

2. Unsolved features in the case of the Pöschl-Teller problem

First, let us consider the Hamiltonian of a particle in a modified Pöschl-Teller potential, whose time-independent Schrödinger equation is [3, 4, 5, 6]

$$H_{m+1}\,\psi = \left(-\frac{\hbar}{2m}\frac{d^2}{dx^2} - \frac{\alpha^2 m(m+1)}{\cosh^2 \alpha x}\right)\psi = E\,\psi\;, \tag{1}$$

where $\alpha > 0$, and the integer m is greater than 0. To shorten the algebraic equations we shall set $\frac{-\hbar}{2m} = 1$.

The eigenvalue problem may be solved using the Infeld & Hull's (IH) factorizations,[7]

$$A^+_{m+1} A^-_{m+1}\, \psi^m_\ell = (H_{m+1} + \epsilon_{m+1})\, \psi^m_\ell, \tag{2a}$$

$$A^-_m A^+_m\, \psi^m_\ell = (H_{m+1} + \epsilon_m)\, \psi^m_\ell, \tag{2b}$$

where the raising/lowering operators are given by

$$A^{\mp}_m = \alpha m \tanh \alpha x \mp \frac{d}{dx}\,, \tag{3}$$

where $\epsilon_m = \alpha^2 m^2$, and $\ell = m - n$, where n is the eigenvalue index,

$$E_n \equiv E_{m,n} = -\alpha^2 (m-n)^2, \qquad\qquad n = 0, 1, 2... < m. \tag{4}$$

The coefficient out of the parenthesis in eq.(3) is only used to normalize the eigenfunctions, and we shall drop it here on.

Note that we have changed the usual notation for the eigenvalues in this problem, since we shall show below that there are some hidden features masked by the shorter notation. To do so, let us remember that the IH operators change the upper index of the eigenfunctions; in fact, the eigenfunctions $\psi_n(x) \equiv \psi^m_{m-n}(x)$ have to be be found by succesive applications of the raising operator on the zeroth order eigenfunctions defined by

$$\psi_0 \equiv \psi^\ell_\ell(x) = \sqrt{\frac{\alpha\Gamma(\ell + \frac{1}{2})}{\sqrt{\pi}\Gamma(\ell)}} \cosh^{-\ell} \alpha x. \tag{5}$$

leaving the lower index ℓ unchanged, given that, up to a constant factor,

$$A^-_{s+1}\, \psi^s_\ell = \psi^{s+1}_\ell, \quad A^+_s\, \psi^s_\ell = \psi^{s-1}_\ell. \tag{6}$$

We would like to stress that in the usual SUSY treatment of this problem,[8] it seems that the following fact was not taken under consideration: if we look at the factorizations (2), we notice that interchanging the order of the IH operators seem to factorize the same hamiltonian, but only if the index of the raising/lowering operators is changed! A simple change of order produces another physical problem, with a different potential function. This becomes evident if we do the following algebra:

$$A^-_m A^+_m\, \psi^m_{m-n} = (H_{m+1} + \epsilon_m)\, \psi^m_{m-n} = (E_{m,n} + \epsilon_m)\, \psi^m_{m-n},$$

but also, due to (6) and (2a),

$$A^-_m A^+_m\, \left(A^-_m \psi^{m-1}_{m-n}\right) = A^-_m\, \left(A^+_m A^-_m \psi^{m-1}_{(m-1)-(n-1)}\right) = \tag{7a}$$

$$A^-_m\, (H_m + \epsilon_m)\, \psi^{m-1}_{(m-1)-(n-1)} = (E_{m-1,n-1} + \epsilon_m)\, A^-_m \psi^{m-1}_{m-n} = (E_{m,n} + \epsilon_m)\, \psi^m_{m-n} \tag{7b}$$

where we used our notation to emphasise that in eq.(7a) the problem in parenthesis refers to a different potential than the problem in the preceding equation, however rendering the same energy eigenvalue! [9] Therefore, we think that it may be possible that the usual SUSY approach to this problem has to be reevaluated.[10]

3. Dawson's function

Let us now consider Dawson's function, which may be defined in terms of the error function. Dawson's function is directly related to the error function of imaginary argument [11]

$$F(x) = e^{-x^2} \int_0^x e^{y^2}\, dy = -\frac{\sqrt{\pi}\, i}{2}\, e^{-x^2} \operatorname{erf}(ix) \tag{8}$$

and its properties are better determined in terms of its derivatives,

$$F'(x) = -2xF(x) + 1 \tag{9a}$$

$$F^{(k+1)}(x) + 2xF^{(k)}(x) + 2kF^{(k-1)}(x) = 0\,. \tag{9b}$$

Attempts have been made to find a simple representation of this function in terms of elementary functions, for example, that of Cody *et al.*,[12] who proposed a representation of $F(x)$ in terms of rational approximations.

We shall show here that there exists a singular SUSY type relation between Dawson's function and its derivatives, to the eigenfunctions of the wrong-sign Hermite differential equation, which is found in a new factorization of the Hamiltonian of the simple harmonic oscillator (SHO).[13]

3.1. *Generalized factorization of the SHO Hamiltonian*

In a previous article [13] we have developed two factorizations of the SHO Hamiltonian in terms of two non-selfadjoint operators

$$B^- = \frac{1}{\sqrt{2}}\left(\alpha^{-1}(x)\frac{d}{dx} + \beta(x)\right), \tag{10a}$$

$$B^+ = \frac{1}{\sqrt{2}}\left(-\alpha(x)\frac{d}{dx} + \beta(x)\right). \tag{10b}$$

In the first factorization, we required that $B^-B^+ = H + \frac{1}{2}$. Upon inverting the product, B^+B^-, the two parameter solutions for $\alpha(x)$ and $\beta(x)$ defined a Sturm-Liouville equation which included the quantum mechanics SHO equation, its SUSY partners,[1] and Hermite's equation, as particular cases for defined regions of the two-parameter space.[13]

In a second factorization, we proposed that the Hamiltonian be factorized as $B^+B^- = H - \frac{1}{2}$, which is possible if now the functions $\alpha(x)$ and $\beta(x)$ depend on a single parameter, γ_3, and are given by

$$\alpha_{\gamma_3}(x) = \sqrt{1 + \gamma_3 e^{x^2}}\,, \qquad \beta_{\gamma_3}(x) = \frac{x}{\sqrt{1 + \gamma_3 e^{x^2}}}. \tag{11}$$

The inverse operator product B^-B^+ now defines a new eigenvalue equation

$$\mathcal{L}_{\gamma_3}\mathrm{H}_n^{\gamma_3} + \lambda_n \omega_{\gamma_3}(x)\mathrm{H}_n^{\gamma_3} = 0, \tag{12}$$

where

$$\mathcal{L}_{\gamma_3} = \left(1 + \gamma_3 e^{x^2}\right)\frac{d^2}{dx^2} + 2\gamma_3 x e^{x^2}\frac{d}{dx} + \frac{\gamma_3 e^{x^2} + \gamma_3^2 e^{2x^2} - x^2}{1 + \gamma_3 e^{x^2}} \tag{13}$$

is a one-parameter self-adjoint operator with the weight function $\omega_{\gamma_3}(x) = 2\left(1 + \gamma_3 e^{x^2}\right)$, and it is isospectral to the quantum SHO Hamiltonian, which is obtained in the limit $\gamma_3 \to 0$. The eigenfunctions in this case are

$$\mathrm{H}_n^{\gamma_3}(x) = B^-\psi_{n+1}(x)\,. \tag{14}$$

where $\psi_n(x)$ are the SHO eigenfunctions.

For this work, it is very interesting to note that in the large limit $\gamma_3 \gg 1$, one can obtain the wrong-sign Hermite's differential equation

$$\left[\frac{d^2}{dx^2} + 2x\frac{d}{dx} + 2(n+1)\right] \widetilde{H}_n(x) = 0 \; , \tag{15}$$

which differs from the Hermite equation only in the sign in front of the first derivative. The corresponding eigenfunctions are of the quantum oscillator type, but vanishing faster due to a squared exponential factor, $\widetilde{H}_n(x) = e^{-x^2} H_n(x)$.

From the Hermite polynomials' recursion relations it is easy to find the raising and lowering operators for these functions. They turn out to be

$$\left(\frac{d}{dx} + 2x\right) \widetilde{H}_n(x) = 2n\widetilde{H}_{n-1}(x), \tag{16a}$$

$$-\frac{d}{dx}\widetilde{H}_n(x) = \widetilde{H}_{n+1}(x). \tag{16b}$$

Note that the reversed sign in the first derivative term of eq.(15) produces these "reversed" Hermite polynomials' recursion relations.

3.2. *A singular SUSY relation and Dawson's eigenfunctions*

As one can see, equation (15) is the same as (9b) when $k = 1$ and $n = 0$, i.e., the ground state equation of the wrong-sign Hermite eigenvalue problem, whose eigenfunction is just the Gaussian function

$$\widetilde{H}_0(x) = e^{-x^2} \tag{17}$$

and it is not Dawson's function! Now, $\widetilde{H}_0(x)$ and $\mathrm{F}(x)$ are completely different, and their first derivatives are also different,

$$\widetilde{H}_0'(x) = -2x\widetilde{H}_0(x) \; , \tag{18a}$$

$$\mathrm{F}'(x) = -2x\mathrm{F}(x) + 1 \, , \tag{18b}$$

however, they share the same second order differential equation. The reason for this is that Dawson's function is the second solution of equation (15) when $n = 0$, since starting with $\widetilde{H}_0(x)$, that solution is [14]

$$f(x) = \widetilde{H}_0(x) \int_0^x \frac{e^{-\int^y 2z\,dz}}{\left(\widetilde{H}_0(y)\right)^2} dy = F(x) \tag{19}$$

It is even more interesting to see that, using eqs.(9), (15) and (16), we can find a SUSY like ladder relation between the wrong-sign Hermite eigenfunctions $\widetilde{H}_n(x)$, and the derivatives of Dawson's functions, which we shall hereafter call the Dawson's eigenfunctions, $D_n(x) \equiv F^{(n-1)}(x)$, for $n = 1, 2, 3, \ldots$, and hence $D_1(x) \equiv F(x)$. This ladder relation is shown in Fig.1.

The fact that $F(x)$ has one zero at $x = 0$, while the Gaussian does not have any zero for all finite x, implies that in order that both sets of eigenfunctions cover the whole space of non-singular functions $f(x)$, there must exist an additional function $D_0(x)$, however enlarging the eigenvalue equations for Dawson's eigenfunctions. This feature is the equivalent to the SUSY-QM procedure, where the zero-th order SUSY partner eigenfunction is missing. To find the missing $D_0(x)$, we can see that eq.(9), in the case $k = 0$, has two solutions, the constant solution, and the error function. However, if we assume that the recursion relation (16a) gives rise to eq.(9a), then, the (non-trivial) zero-th eigenfunction ought to be $D_0(x) = const.$

We say here that Dawson's eigenfunctions and the wrong-sign Hermite eigenfunctions possess a singular SUSY relation, although (*i*) there is no QM or any type of physics problem associated, (*ii*) their relation did not arise from an operator procedure; moreover, they share the recurrence relations and the second order differential equation, (*iii*) there does not exist an associated SUSY parameter in this relation, and (*iv*) in order to cover the space of nonsingular functions of x, there must exist a zero-less eigenfunction $D_0(x)$, which in this case is the constant function.

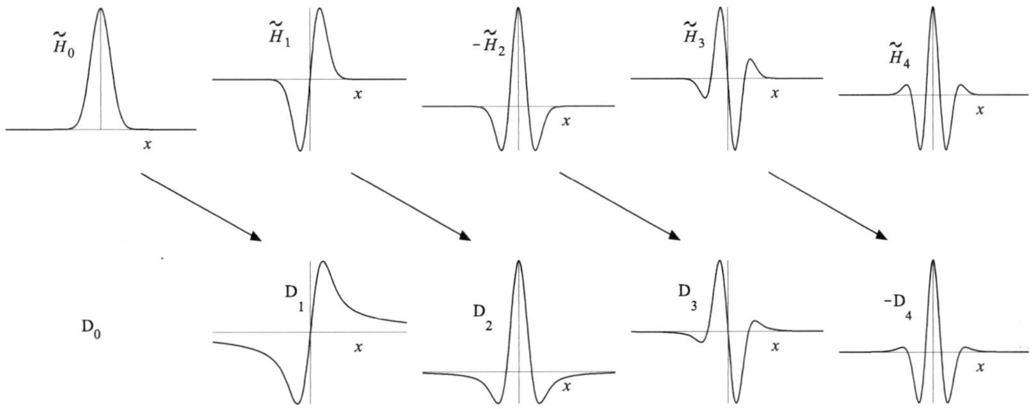

Figure 1. The SUSY type ladder relation between Dawson's eigenfunctions $D_n(x)$ and the wrong-sign Hermite eigenfunctions $\widetilde{H}_n(x)$. The function $D_0(x)$ is missing, but can be found using the second order differential equation (9b) and the lowering operator (9a).

4. Conclusions

We have tried to motivate here that some aspects of the SUSY algebraic method still need to be considered, and some are hidden in known algebra, awaiting for a deeper or broadened approach in the field. The case of the Pöschl-Teller problem shows that we must reconsider some work already done in the area, while the relation of Dawson's function to the wrong-sign Hermite eigenfunctions shows that some hidden SUSY aspects are still awaiting to be found, which may extend the SUSY techniques to other areas in mathematical physics.

References

[1] Mielnik B 1984 *J. Math. Phys.* **25** 3387
[2] García-Daz J S CITAR *J. Math. Phys.* **25** 3387
[3] Pöschl G. and Teller E 1933 *Zeitschrift für Physik* **83** 143
[4] Rosen N and Morse P M 1932 *Phys. Rev.* **42** 210
[5] Cooper F, Khare A and Sukhatme U 2001 *Supersymmetry in Quantum Mechanics* (Singapore: Scientific)
[6] Siegfried F 1994 *Practical Quantum Mechanics*, (Berlin: Springer-Verlag)
[7] Infeld L and Hull T E 1951 *Rev. Mod. Phys.* **23** 21
[8] Díaz J I, Negro J, Nieto L M and Rosas-Ortiz O 1999 *J. Phys. A: Math. Gen.* **32** 8447
[9] We think this is the reason for the comment of the authors in the .., paragraph in page ....
[10] Arcos-Olalla R 2017 *Ph.D. Thesis, University of Guanajuato, México*
[11] Sajo E 1993 *J. Phys. A: Math. Gen.* **26** 2977
[12] Cody W J, Paciorek Kathleen A and Thacher H C 1970 *Math. Comp.* **24** 171
[13] Arcos-Olalla R, Reyes M A and Rosu H C 2012 *Physics Letters A* **376** 2860
[14] Arfken G B and Weber H J 2001 *Mathematical methods for physicists* (Elsevier) p 536-537

Institute of Physics
Dirac House, Temple Back
Bristol BS1 6BE UK

ISSN: 1742-6588
ISBN 978-1-5108-6490-0

CURRAN ASSOCIATES INC.
proceedings.com